CH00764664

THE MERLIN 100 SERIES

– the ultimate military development

Alec Harvey-Bailey and Dave Piggott

HISTORICAL SERIES No. 19

Published in 1993 by the
Rolls-Royce Heritage Trust
PO Box 31 Derby England

ISBN: 1-872922-04-X

The Historical Series is published as a joint initiative by the
Rolls-Royce Heritage Trust and The Sir Henry Royce Memorial Foundation

Cover Picture: Pat Fillingham, de Havilland test pilot in a Sea Hornet F Mk 20
        (Photos BAe via Winged Memories)

Printed by Bemrose Security Printing, Derby

THE MERLIN 100 SERIES
– the ultimate military development

CONTENTS

FOREWORD

This book is intended to form an extension of "The Merlin in Perspective - the combat years" and also to complement A. A. Rubbra's "Rolls-Royce Piston Aero Engines - a designer remembers". Small differences may exist in information given in this book as a result of further research and should be regarded as being more accurate.

The illustrations which form the bulk of the book owe their presence to three people, Cyril Lovesey, Tony Dunwell and David Piggott. Cyril Lovesey (Lov) was Chief Development Engineer of the Merlin engine during its most important phase. He saw the need to 'sell' major changes, not only to the customers in the form of Ministries and Fighting Services, but also to the rest of the Company who would have to implement and support the product. He engaged Tony Dunwell (Lov/Dnl) to produce brochures illustrating new features, providing both enhanced performance and cures to existing problems, in a way that was easily assimilable to all concerned. Tony Dunwell established the Technical Illustration Section as a valuable part of the engineering organisation by virtue of blending artistic skills with technical details. The ensuing brochures were widely appreciated and as one retired senior engineer has said they enabled one part of Development to see what the others were doing. The Technical Illustration section still exists in Rolls-Royce plc and its work is much respected by other engine manufacturers. It is thanks to David Piggott's efforts that the Dunwell drawings, together with other information, have been found in Company Archives, thus enabling this book to be produced.

Alec Harvey-Bailey

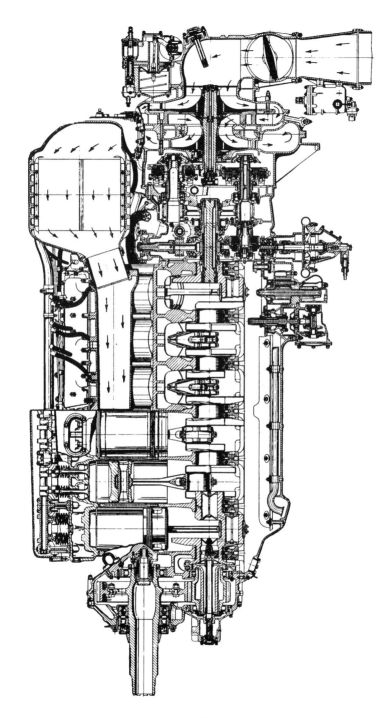

Rolls-Royce Merlin 100 Series Aero Engine

THE MERLIN 100 SERIES
– the ultimate military development

The 100 series Merlin came about not as the result of a 'clean sheet of paper' redesign, but the concentration of features shown to be desirable from hard-won combat experience, plus new requirements to meet planned horsepower and altitude performance developments.

Central to the engine was the decision to go for end feed crankshaft lubrication affecting crankshaft, crankcase, wheelcase and reduction gear. It would also be an opportune moment to embody a two-stage supercharger with an overhung first-stage impeller and an improved intake elbow, resulting from the adoption of single-point fuel injection. A further change, easily applicable to earlier engines, was a double packless gland ball bearing coolant pump. With such alterations to major components it was possible to embody other less extensive changes to enable the use of 30lb/sq.in. combat boost on certain marks (for those more familiar with boost measurements such as bar, atmospheres and inches of mercury absolute Rolls-Royce boost measurements were in lb/sq. in. (gauge), thus one has to add 14.7lb to the indicated boost before converting to other notations).

Comment has been made on the choice of single-point fuel injection. If one goes back to the Battle of Britain there was criticism of the S.U. carburettor, because of an engine cut-off for approximately 1.5 seconds when entering a dive under negative-g conditions. This was largely overcome with the anti-g version of the S.U., while certain marks adopted the Bendix carburettor, which was of the low pressure injection type. For the 100 series, single-point injection was chosen because of its accuracy in fuel metering and its complete freedom from any g effects. By injecting fuel into the eye of the overhung impeller excellent atomisation was achieved, which with the latent heat effect in the supercharger reduced charge temperature by 20°C, an advantage lost on direct injection engines as used by the enemy. It also reduced the number of fuel pipes on the engine.

The first operational use of Merlin 100 series engines was in early 1944, when to take advantage of improved altitude performance five Mosquito PR 32 aircraft were built by de Havillands, with extended wings and 113/114 engines built by the Experimental Department. Experience with these aircraft, flying from RAF Benson, showed a tendency to supercharger surge at high altitude.

This, including flying from Hucknall, led to the introduction of Merlin 113A/114A with anti-surge diffusers, having slightly reduced throat areas. Installed in PR34 Mosquitoes they saw squadron service before the end of the European war. Bomber B35 and fighter NF36 types were later reaching service, but remained on the squadrons until superseded by jet powered aircraft in the early 1950s.

The single-seater de Havilland Hornet long-range fighter was powered by 130 series engines. Changes from the 113/114 included handed rotation, the engines being fitted in pairs, while a side mounted main coolant pump and down draught air intake reduced frontal area. These engines, rated at 25lb boost, had development potential for + 30lb. Over 200 aircraft were delivered to the RAF and the modified

Sea Hornet was used by the Fleet Air Arm. The Hornet was renowned for its performance and handling, but its initial appearance was overshadowed by the publicity given to the new jet fighters. Its ability to give the early marks of Meteor a hard time at altitude is rarely, if ever, mentioned and this brilliant aeroplane passed out of RAF service in 1955 with no survivor. The Spitfire and Hurricane will be remembered for their role in the Battle of Britain and in the mind's eye there will always be the blue English summer skies, criss crossed by vapour trails, visible signs of an epic struggle. The Hornet has no such history, apart from some grim operations against terrorists in Malaya.

The Mosquito and Hornet were the only aircraft in Britain to make real use of the 100 series engine, but in America, Packard Merlin 100s, the V1650-9 and V1650-11, were used by the U.S. Army Air Force in later Mustangs. Today one comes across historic aircraft, including Spitfires, with V1650-9s. The racing Mustangs in America use hybridised Merlins to suit particular conditions.

Towards the end of the war the Australian Government had taken a licence to build the Lincoln B30 bomber. 73 aircraft were built, some of which had the fuselage lengthened to suit a maritime role.

The first 50 had Merlin 85s but the remaining 23 were fitted with Merlin 102s built under licence by Commonwealth Aircraft Corporation at their Lidcombe NSW, factory. Some earlier aircraft may also have been converted to 102s. In Britain there were also short production runs of Merlin 102s and 104s. The former had an auxiliary gearbox drive, while the latter had engine mounted accessories and was intended for Mosquito aircraft with emphasis on lower level performance. The Merlin 104 was not proceeded with, but some Merlin 102s were used for development flying. These included early flying on Tudor I, whose first flight was in June 1945 and the Tudor II in January 1946. A BOAC Lancastrian G-AGJI converted from a Lancaster I was fitted with Merlin 102s in May 1946, while Lincoln RE258 with Merlin 102s in the in-board power plants took part in the Canadian winterisation trials of 1946-1947. A project for the Fleet Air Arm, the Short Sturgeon, was fitted with Merlin 140s with contra-rotating reduction gears, but the aircraft did not reach squadron service. The final development for the Hornet consisted of Merlin 134/135s which were fitted with Corliss throttles to give less throttle torque and a clean intake at full throttle conditions. The Argentine Government had a project for the Nancu fighter, using 134/135s which came to nothing.

Engine programmes and performance curves are shown in the text, but mention must be made of a high power development at the RM17SM rating. This featured larger diameter supercharger rotors and a longer duration exhaust cam with increased overlap. Engine 90369 carried out a ten hour flight approval test of which 30 minutes was completed at 2340 hp at 3000 rpm and 30lb/sq. in. boost, illustrating the military potential of the design.

The Dunwell illustrations show features in the initial development engines as well as those released to Production and naturally a number of items changed as development continued.

Merlin 100 series formed the basis for the 600 and 700 series commercial engines. With the virtual demise of the Tudor the only user was Canadair DC4M, particularly in its -2 and -4 form. Initially many problems were encountered. Whilst some of these could be attributed to taking a highly developed military engine where further development action was limited for commercial operation, they could have been better managed and controlled had there been more experience of the demands of civil operation within the Company. As it was, both the engine and Company personnel were on a steep learner curve, which at times was not helped by the customer. Ultimately a good standard of operation was achieved and laid the foundation of a more mature Company approach to the Airline business, as commented on in "Hives, The Quiet Tiger" and "The Sons of Martha".

THE DUNWELL DRAWINGS

Lov used Tony Dunwell's drawings to sell his ideas throughout the development of the 100 series and in this, the main body of the book, we show examples from each stage.

The first section dealing with the projected developments and dating from the Spring of 1943 shows the early specification of the engine. The next section showing the features of the 100 series, includes illustrations of the first batch of experimentally built engines from the autumn of 1943, similar drawing of the initial production standard of engines from the spring of 1944 and drawings from the winter of 1944/45 which show some of the changes introduced into the production engine specification derived from the ongoing development programme.

The final two sections show drawings of the 130/131 series of engine, again from the winter of 1944/45.

Frequent references are made to design schemes (e.g. DES 14716) which define the basic engineering definition of a particular feature; DES standing for Derby Engine Scheme, the register for which opened in November 1923 and continued throughout the piston engine era. Other references are made to detail part numbers and these are shown, for example, simply as D24381. The DES schemes are still kept in the drawing stores at East Kilbride, Scotland, whilst the Merlin detail drawings are in the custody of the Heritage Trust at Hucknall.

11

PROJECTED DEVELOPMENT
OF THE
MERLIN
TWO STAGE SUPERCHARGED
ENGINE
DURING THE PERIOD
1943/1944

Title page of a Rolls-Royce brochure – dated 3 November 1942

Projected Development of Merlin Two Stage Intercooled Engine During the Period - 1943/1944.

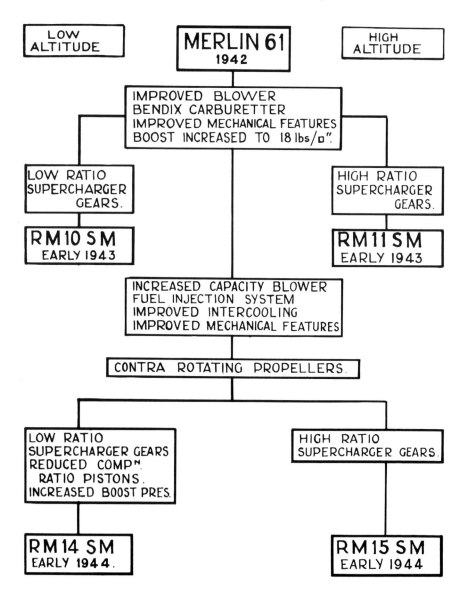

Development strategy in November 1942

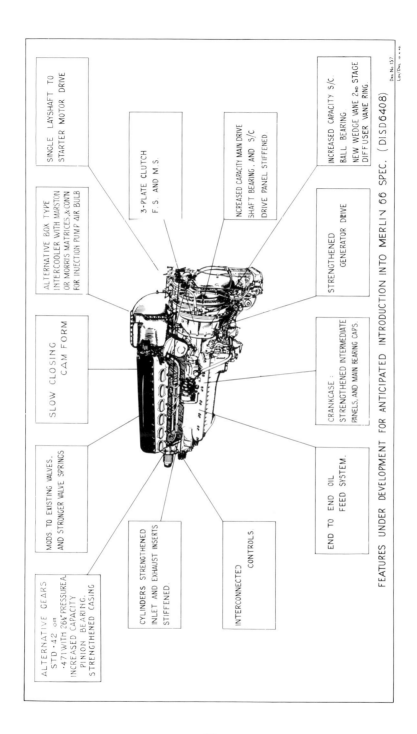

SINGLE LAYSHAFT TO STARTER MOTOR DRIVE

3-PLATE CLUTCH F.S. AND M.S.

INCREASED CAPACITY MAIN DRIVE SHAFT BEARING, AND S/C DRIVE PANEL STIFFENED

INCREASED CAPACITY S/C BALL BEARING NEW WEDGE VANE 2ND STAGE DIFFUSER VANE RING

ALTERNATIVE BOX TYPE INTERCOOLER WITH MARSTON OR MORRIS MATRICES, & CONN'N FOR INJECTION PUMP AIR BULB

STRENGTHENED GENERATOR DRIVE

SLOW CLOSING CAM FORM

CRANKCASE: STRENGTHENED INTERMEDIATE PANELS, AND MAIN BEARING CAPS.

MODS TO EXISTING VALVES. AND STRONGER VALVE SPRINGS

END TO END OIL FEED SYSTEM.

ALTERNATIVE GEARS STD ·42 OR ·471 WITH 26½°PRESSURE A. INCREASED CAPACITY PINION BEARING. STRENGTHENED CASING

CYLINDERS STRENGTHENED INLET AND EXHAUST INSERTS STIFFENED.

INTERCONNECTED CONTROLS

FEATURES UNDER DEVELOPMENT FOR ANTICIPATED INTRODUCTION INTO MERLIN 66 SPEC. (DISD6408)

DRG. No. 157

LOV/DRL. 10 · 2 · 43

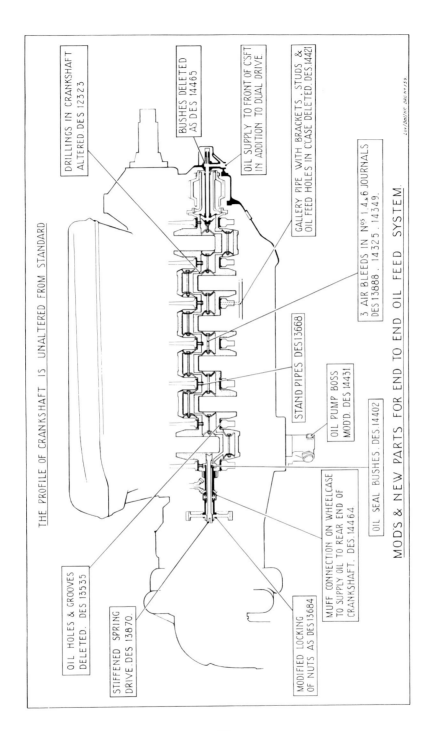

THE PROFILE OF CRANKSHAFT IS UNALTERED FROM STANDARD

DRILLINGS IN CRANKSHAFT ALTERED DES 12323

BUSHES DELETED AS DES 14465

OIL SUPPLY TO FRONT OF CSFT IN ADDITION TO DUAL DRIVE

GALLERY PIPE WITH BRACKETS, STUDS & OIL FEED HOLES IN CCASE DELETED DES 14421

3 AIR BLEEDS IN Nos 1, 4, & 6 JOURNALS DES 13888 . 14325 . 14349.

STAND PIPES DES 13668

OIL PUMP BOSS MODD. DES 14431

OIL SEAL BUSHES DES 14402

MUFF CONNECTION ON WHEELCASE TO SUPPLY OIL TO REAR END OF CRANKSHAFT. DES 14464

MODIFIED LOCKING OF NUTS AS DES 13684

STIFFENED SPRING DRIVE. DES 13870.

OIL HOLES & GROOVES DELETED. DES 15535

Lov/Dmi/HP. DRG No 139

MOD'S & NEW PARTS FOR END TO END OIL FEED SYSTEM.

16

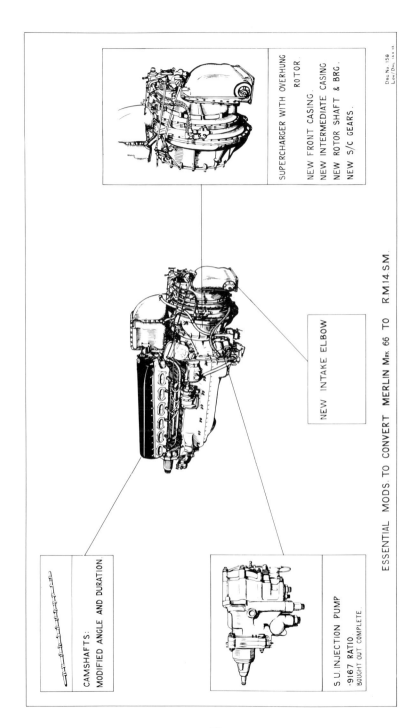

SUPERCHARGER WITH OVERHUNG ROTOR.

NEW FRONT CASING.
NEW INTERMEDIATE CASING
NEW ROTOR SHAFT & BRG.
NEW S/C GEARS.

DRG No. 158
LOV/DML 144 13

NEW INTAKE ELBOW

CAMSHAFTS:
MODIFIED ANGLE AND DURATION

S.U. INJECTION PUMP
·9167 RATIO
BOUGHT OUT COMPLETE.

ESSENTIAL MODS. TO CONVERT MERLIN Mᴋ 66 TO R.M.14.S.M.

17

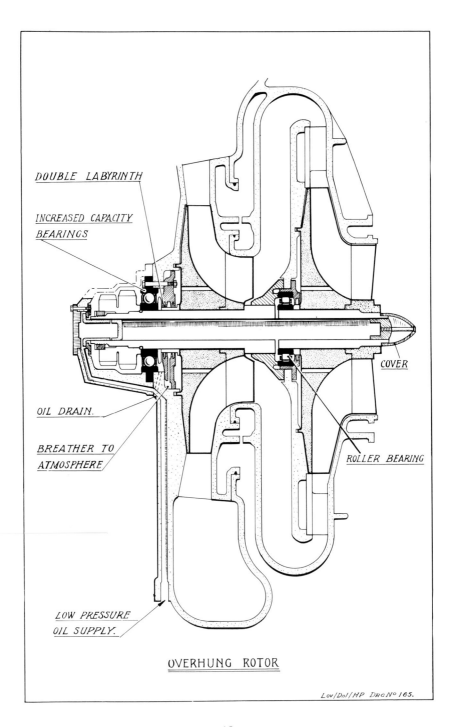

DOUBLE LABYRINTH

INCREASED CAPACITY
BEARINGS

COVER

OIL DRAIN.

BREATHER TO
ATMOSPHERE

ROLLER BEARING

LOW PRESSURE
OIL SUPPLY.

OVERHUNG ROTOR

Lov/Dnl/HP Drg N° 165.

ILLUSTRATED NEW FEATURES
OF THE
MERLIN MK. 100
101, 110, 113, 112 & 114

ISSUED BY Lov/Dnl
TECHNICAL ILLUSTRATION SECTION
ROLLS-ROYCE, DERBY

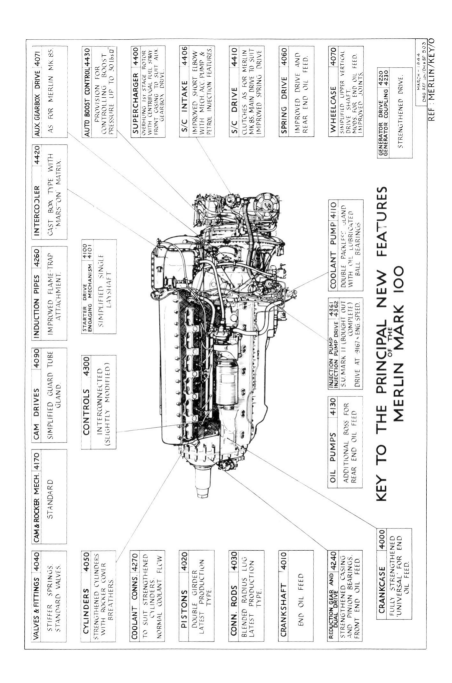

KEY TO THE PRINCIPAL NEW FEATURES OF THE MERLIN MARK 100

VALVES & FITTINGS 4040 — STIFFER SPRINGS. STANDARD VALVES.

CAM & ROCKER MECH. 4170 — STANDARD

CAM DRIVES 4090 — SIMPLIFIED GUARD TUBE GLAND.

INDUCTION PIPES 4260 — IMPROVED FLAME-TRAP ATTACHMENT.

INTERCOOLER 4420 — CAST BOX TYPE WITH MARSTON MATRIX.

AUX GEARBOX DRIVE 4071 — AS FOR MERLIN MK 85.

AUTO BOOST CONTROL 4430 — PROVISION FOR CONTROLLING BOOST PRESSURE UP TO 30 lbsq"

SUPERCHARGER 4400 — OVERHUNG 1st STAGE ROTOR WITH CENTRIFUGAL FUEL SPRAY FRONT CASING TO SUIT AUX GEARBOX DRIVE.

S/C INTAKE 4406 — IMPROVED SHORT ELBOW WITH MECH. A.C. PUMP & PETROL INJECTION FEATURES.

S/C DRIVE 4410 — CLUTCHES AS FOR MERLIN MK 85. MAIN DRIVE TO SUIT IMPROVED SPRING DRIVE

SPRING DRIVE 4060 — IMPROVED DRIVE AND REAR END OIL FEED.

WHEELCASE 4070 — SIMPLIFIED UPPER VERTICAL DRIVE SHAFT. MODS FOR END OIL FEED. IMPROVED JOINTS.

GENERATOR DRIVE 4220 / GENERATOR COUPLING 4230 — STRENGTHENED DRIVE.

CONTROLS 4300 — INTERCONNECTED (SLIGHTLY MODIFIED)

STARTER DRIVE ENGAGING MECHANISM 4100 4101 — SIMPLIFIED SINGLE LAYSHAFT

COOLANT PUMP 4110 — DOUBLE PACKLESS GLAND WITH OIL LUBRICATED BALL BEARINGS.

CYLINDERS 4050 — STRENGTHENED CYLINDERS WITH ROCKER COVER BREATHERS.

COOLANT CONNS. 4270 — TO SUIT STRENGTHENED CYLINDERS. NORMAL COOLANT FLOW

PISTONS 4020 — DOUBLE GIRDER LATEST PRODUCTION TYPE

CONN. RODS 4030 — BLENDED RADIUS LUG LATEST PRODUCTION TYPE.

CRANKSHAFT 4010 — END OIL FEED

INJECTION PUMP 4361 / INJECTION PUMP DRIVE 4362 — S.U. MARK II (BOUGHT OUT COMPLETE) DRIVE AT .9167 x ENG SPEED.

OIL PUMPS 4130 — ADDITIONAL BOSS FOR REAR END OIL FEED

REDUCTION GEAR AND DUAL DRIVE 4240 — STRENGTHENED CASING AND PINION BEARINGS. FRONT END OIL FEED.

CRANKCASE 4000 — FULLY STRENGTHENED 'UNIVERSAL' FOR END OIL FEED.

MARCH - 1944
DRG REF No Merlin KEY/0

REF : MERLIN/KEY/0

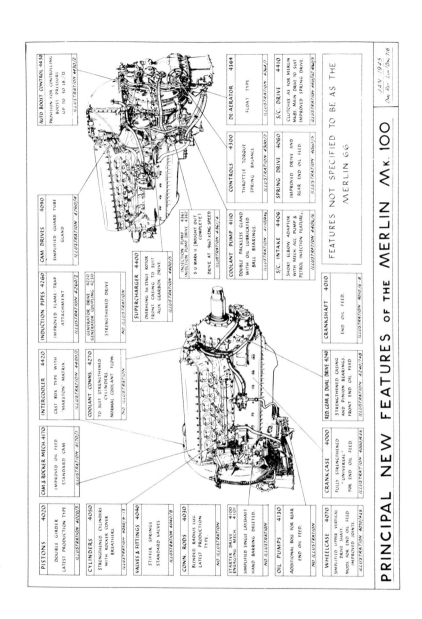

PRINCIPAL NEW FEATURES of the MERLIN Mk. 100

PISTONS 4020
DOUBLE GIRDER
LATEST PRODUCTION TYPE
ILLUSTRATION 4020/3

CYLINDERS 4050
STRENGTHENED CYLINDERS
WITH ROCKER COVER
BREATHERS.
ILLUSTRATION 4050/9-13

VALVES & FITTINGS 4040
STIFFER SPRINGS
STANDARD VALVES
ILLUSTRATION 4040/8

CONN. RODS 4030
BLENDED RADIUS LUG
LATEST PRODUCTION
TYPE.
NO ILLUSTRATION

STARTER DRIVE 4100 / ENGAGING MECH. 4101
SIMPLIFIED SINGLE LAYSHAFT
HAND BARRING DELETED.
NO ILLUSTRATION

OIL PUMPS 4130
ADDITIONAL BOSS FOR REAR
END OIL FEED.
NO ILLUSTRATION

WHEELCASE 4070
SIMPLIFIED UPPER VERTICAL
DRIVE SHAFT.
MODS FOR END OIL FEED
IMPROVED JOINTS
ILLUSTRATION 4070/4&5

CRANKCASE 4000
FULLY STRENGTHENED
"UNIVERSAL"
FOR END OIL FEED
ILLUSTRATION 4000/4&5

CAM & ROCKER MECH. 4170
IMPROVED OIL FEED
STANDARD CAM
ILLUSTRATION 4170/1

INTERCOOLER 4420
CAST BOX TYPE WITH
'MARSTON' MATRIX
ILLUSTRATION 4420/2

COOLANT CONNS. 4270
TO SUIT STRENGTHENED
CYLINDERS
NORMAL COOLANT FLOW.
NO ILLUSTRATION

RED. GEAR & DUAL DRIVE 4240
STRENGTHENED CASING
AND PINION BEARINGS.
FRONT END OIL FEED
ILLUSTRATION 4240/7&8

CRANKSHAFT 4010
END OIL FEED.
ILLUSTRATION 4010/6-8

INDUCTION PIPES 4260
IMPROVED FLAME-TRAP
ATTACHMENT
ILLUSTRATION 4260/2

GENERATOR DRIVE 4220 / GENERATOR COUPLING 4230
STRENGTHENED DRIVE
NO ILLUSTRATION

SUPERCHARGER 4400
OVERHUNG 1st STAGE ROTOR
FRONT CASING TO SUIT
AUX GEARBOX DRIVE
ILLUSTRATION 4400/3

INJECTION PUMP 4361 / INJECTION PUMP DRIVE 4362
S.U. MARK II (BOUGHT OUT
COMPLETE)
DRIVE AT 9167 x ENG SPEED
ILLUSTRATION 4361/4

CAM DRIVES 4090
SIMPLIFIED GUARD TUBE
GLAND
ILLUSTRATION 4260/4

COOLANT PUMP 4110
DOUBLE PACKLESS GLAND
WITH OIL LUBRICATED
BALL BEARINGS.
ILLUSTRATION 4110/4&6

S/C INTAKE 4406
SHORT ELBOW ADAPTOR
WITH MECH. A/C PUMP &
PETROL INJECTION FEATURE.
ILLUSTRATION 4406/6

AUTO BOOST CONTROL 4430
PROVISION FOR CONTROLLING
BOOST PRESSURE
UP TO 30 LB./□
ILLUSTRATION 4430/2

CONTROLS 4300
THROTTLE TORQUE
SPRING BALANCE
ILLUSTRATION 4300/2

SPRING DRIVE 4060
IMPROVED DRIVE AND
REAR END OIL FEED.
ILLUSTRATION 4060/6

DE-AERATOR 4364
FLOAT TYPE
ILLUSTRATION 4364/1

S/C DRIVE 4410
CLUTCHES AS FOR MERLIN
Mk85 MAIN DRIVE TO SUIT
IMPROVED SPRING DRIVE.
ILLUSTRATION 4410/2 4410/3

FEATURES NOT SPECIFIED TO BE AS THE
MERLIN 66

JAN 1945
Dwg Ref Lin/Dwn 716

REDUCTION GEAR RATIOS	
Mk 100	.420
Mk 101	.420
Mk 110	.471
Mk 113	.420

S/C	GEAR RATIOS	RATING
Mk 100	5·79 & 7·06	RM 14 SM
Mk 101	5·79 & 7·06	RM 14 SM
Mk 110	6·39 & 8·03	RM 16 SM
Mk 113	6·39 & 8·03	RM 16 SM

AUX. GEARBOX DRIVE

DRIVE FITTED TO Mk 100
PROVISION ONLY ON:-
Mk 101, 110. 113.

COOLANT	FLOW
Mk 100	NORMAL
Mk 101	REVERSED
Mk 110	NORMAL
Mk 113	REVERSED

VARIATIONS of MERLIN Mk 100
WITH 'UNIVERSAL TYPE' C/CASE   MERLIN Mk 101, 110, 113.

MARCH 1944
DRG. REF. Lov/Dnl&BP.478
REF: MERLIN/KEY/1

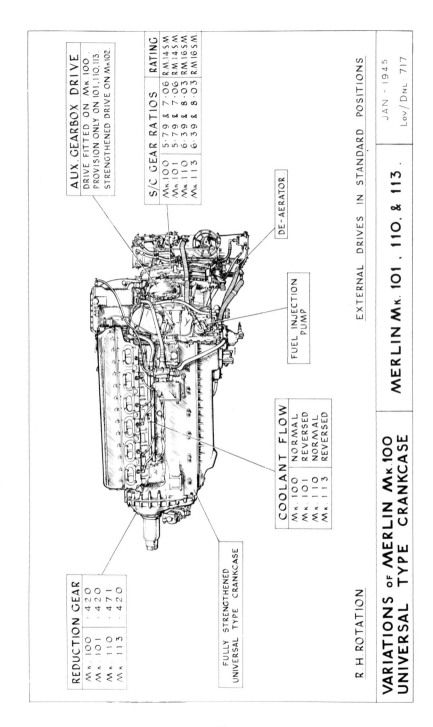

REDUCTION GEAR

Mk. 100	· 4 2 0
Mk. 101	· 4 2 0
Mk. 110	· 4 7 1
Mk. 113	· 4 2 0

FULLY STRENGTHENED
UNIVERSAL TYPE CRANKCASE

COOLANT FLOW

Mk. 100	NORMAL
Mk. 101	REVERSED
Mk. 110	NORMAL
Mk. 113	REVERSED

FUEL INJECTION
PUMP

DE-AERATOR

AUX. GEARBOX DRIVE

DRIVE FITTED ON Mk. 100.
PROVISION ONLY ON 101, 110, 113.
STRENGTHENED DRIVE ON Mk. 102.

S/C GEAR RATIOS | **RATING**

	S/C GEAR RATIOS	RATING
Mk. 100	5·79 & 7·06	R M 14 S M
Mk. 101	5·79 & 7·06	R M 14 S M
Mk. 110	6·39 & 8·03	R M 16 S M
Mk. 113	6·39 & 8·03	R M 16 S M

EXTERNAL DRIVES IN STANDARD POSITIONS

R H ROTATION

VARIATIONS of MERLIN Mk. 100
UNIVERSAL TYPE CRANKCASE

MERLIN Mk. 101, 110, & 113.

JAN - 1945

Lov/Dnl: 717

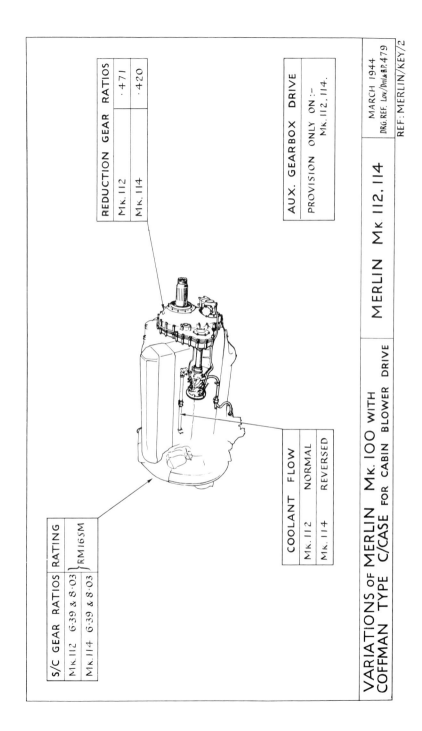

S/C GEAR RATIOS	RATING	
Mk.112	6·39 & 8·03	⎫
Mk.114	6·39 & 8·03	⎬ RM16SM

REDUCTION GEAR RATIOS	
Mk.112	·471
Mk.114	·420

COOLANT FLOW	
Mk.112	NORMAL
Mk.114	REVERSED

AUX. GEARBOX DRIVE
PROVISION ONLY ON :-
Mk.112, 114.

VARIATIONS of MERLIN Mk.100 WITH
COFFMAN TYPE C/CASE FOR CABIN BLOWER DRIVE

MERLIN Mk 112, 114

MARCH 1944
DRG.REF. Low/Dml&B.P.479

REF: MERLIN/KEY/2

COFFMAN TYPE REDUCTION GEAR CASING

TO INCLUDE STRENGTHENING AROUND PINION BEARING HOUSING (TO DES. 15748) AS FOR 'UNIVERSAL' TYPE CASING SPECIFIED FOR MERLIN Mk 100.

NOTE: REDUCTION GEAR RATIO 0·42:1 AS FOR MERLIN Mk 100.

NOTE: THE DESIGN OF THE 'GENERAL PURPOSE' CRANKCASE WHICH IS NOW IN HAND, WILL PROVIDE FOR A CABIN BLOWER DRIVE SUITABLE FOR THIS MARK OF MERLIN.

PROVISION ONLY FOR AUXILIARY GEARBOX DRIVE. BLANKING COVER TO BE FITTED

COFFMAN TYPE CRANKCASE

TO INCLUDE ALL STRENGTHENING AND END OIL FEED FEATURES AS FOR THE 'UNIVERSAL' TYPE CRANKCASE SPECIFIED FOR MERLIN Mk 100

COOLANT CONNECTIONS FOR REVERSE FLOW COOLING TO SUIT MOSQUITO INSTALLATION.

CABIN BLOWER DRIVE FOR 0·42:1 RATIO REDUCTION GEAR AND OIL RELIEF VALVE & FITTINGS TO SUIT COFFMAN TYPE CRANKCASE AS FITTED TO MERLIN Mk 73, 77 etc,

F.S: D 24249   M.S: D 25310

S/C DRIVING GEARS
FOR 6·39 & 8·03 RATIOS.

AS FOR MERLIN Mk. 70,76,77 etc. TO SUIT STARFISH F.S AND THREE PLATE M.S CLUTCHES AS SPECIFIED FOR MERLIN Mk 100

M.S: D 26663   F.S: D 26662

S/C ROTOR SHAFT GEARS
FOR 6·39 & 8·03 RATIOS

TEETH AS FOR MERLIN Mk 70, 76, 77 etc. TO SUIT ROTOR SHAFT OF OVERHUNG ROTOR S/C AS SPECIFIED FOR MERLIN Mk 100

CHANGES TO MERLIN Mk 100 SPECIFICATION TO MAKE MERLIN Mk 114 (RM16 SM)

SUITABLE FOR P.R.U. MOSQUITO AIRCRAFT.

DRG REF: LOV/Dnl 477

REF: MERLIN/KEY/3

26

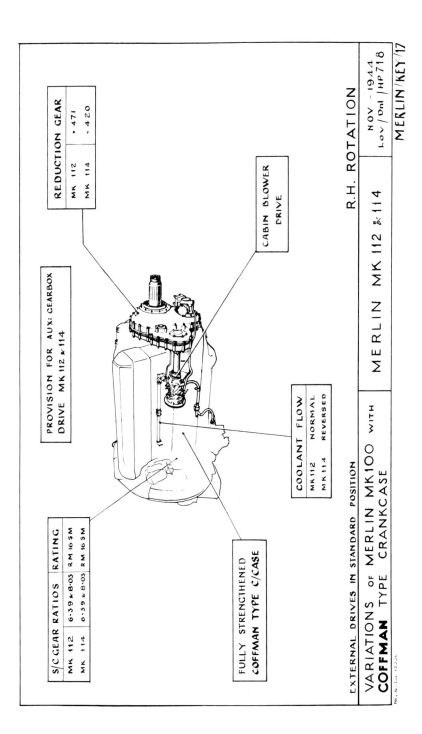

REDUCTION GEAR

| MK 112 | .471 |
| MK 114 | .420 |

PROVISION FOR AUX: GEARBOX
DRIVE MK 112 & 114

CABIN BLOWER
DRIVE

S/C GEAR RATIOS	RATING
MK 112	6·39 & 8·03	RM 16 SM
MK 114	6·39 & 8·03	RM 16 SM

COOLANT FLOW

| MK112 | NORMAL |
| MK 114 | REVERSED |

FULLY STRENGTHENED
COFFMAN TYPE C/CASE

R.H. ROTATION

NOV - 1944
LOV / Onl / HP 718

MERLIN / KEY /17

EXTERNAL DRIVES IN STANDARD POSITION

VARIATIONS OF MERLIN MK100 WITH
COFFMAN TYPE CRANKCASE

MERLIN MK 112 & 114

NEG. No. 13223

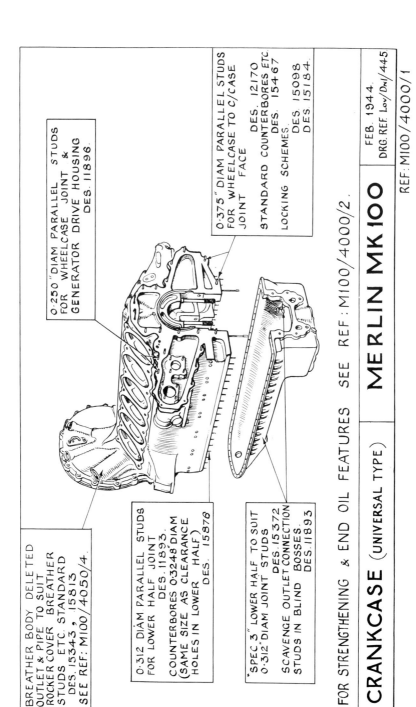

0.250" DIAM PARALLEL STUDS
FOR WHEELCASE JOINT &
GENERATOR DRIVE HOUSING
DES. 11896.

0.375" DIAM PARALLEL STUDS
FOR WHEELCASE TO C/CASE
JOINT FACE   DES. 12170
STANDARD COUNTERBORES ETC
                    DES. 15467
LOCKING SCHEMES :
                    DES. 15098
                    DES. 15184.

BREATHER BODY DELETED
OUTLET & PIPE TO SUIT
ROCKER COVER BREATHER
STUDS ETC. STANDARD
DES. 15343, 15813
SEE REF: M100/4050/4.

0.312 DIAM PARALLEL STUDS
FOR LOWER HALF JOINT
          DES. 11893.
COUNTERBORES 0.3248" DIAM
(SAME SIZE AS CLEARANCE
HOLES IN LOWER HALF)
          DES. 15876

"SPEC 3" LOWER HALF TO SUIT
0.312" DIAM JOINT STUDS
          DES. 15372
SCAVENGE OUTLET CONNECTION
STUDS IN BLIND BOSSES.
          DES. 11893

FOR STRENGTHENING & END OIL FEATURES   SEE   REF: M100/4000/2.

MERLIN MK 100

CRANKCASE (UNIVERSAL TYPE)

FEB. 1944.
DRG. REF. Lov/Dɴᵢ/445

REF: M100/4000/1

28

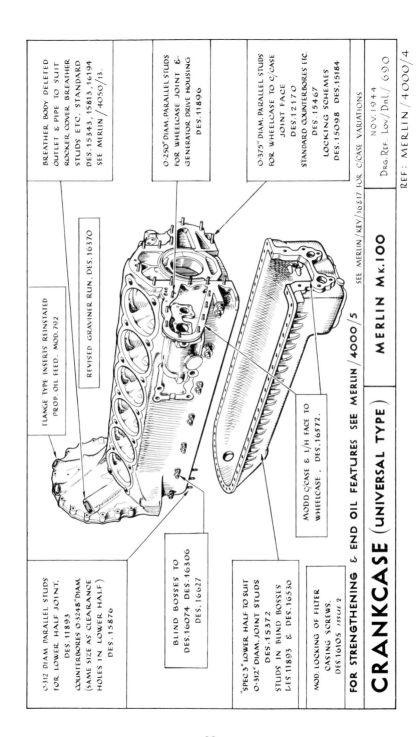

BREATHER BODY DELETED OUTLET & PIPE TO SUIT ROCKER COVER BREATHER STUDS ETC. STANDARD DES.15343, 15813, 16194 SEE MERLIN/4050/13.

0·250" DIAM. PARALLEL STUDS FOR WHEELCASE JOINT & GENERATOR DRIVE HOUSING DES.11896

0·375" DIAM. PARALLEL STUDS FOR WHEELCASE TO C/CASE JOINT FACE DES.12170 STANDARD COUNTERBORES ETC. DES.15467 LOCKING SCHEMES DES.15098 DES.15184

FLANGE TYPE INSERTS REINSTATED PROP. OIL FEED. MOD.792

REVISED GRAVINER RUN. DES.16370

0·312 DIAM PARALLEL STUDS FOR LOWER HALF JOINT. DES.11893 COUNTERBORES 0·3248"DIAM. (SAME SIZE AS' CLEARANCE HOLES IN LOWER HALF) DES.15876

BLIND BOSSES TO DES.16074 DES.16306 DES.16627

'SPEC 3' LOWER HALF TO SUIT 0·312" DIAM. JOINT STUDS DES.15372 STUDS IN BLIND BOSSES LES 11893 & DES.16530

MOD. LOCKING OF FILTER CASING SCREWS. DES.16105 ISSUE 2

MODD C/CASE & L/H FACE TO WHEELCASE. DES.16572.

NOV.1944 DRG.REF. Lov/Dnl./ 690

REF: MERLIN/4000/4

MERLIN Mk.100

CRANKCASE (UNIVERSAL TYPE)

FOR STRENGTHENING & END OIL FEATURES SEE MERLIN/4000/5

SEE MERLIN/KEY/16817 FOR C/CASE VARIATIONS

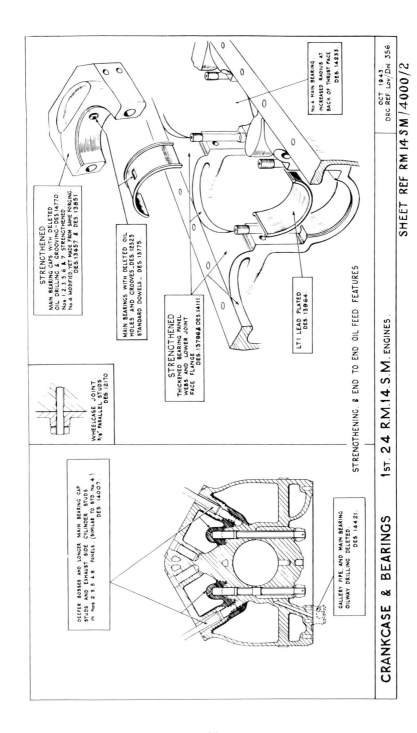

STRENGTHENED
MAIN BEARING CAPS WITH DELETED
OIL DRILLING & GROOVING-DES.14770.
Nos. 1, 2, 3, 5, 6 & 7 STRENGTHENED
No.4 MODIFIED, YET MADE FROM SAME FORGING.
DES.13657 & DES.13851

MAIN BEARINGS WITH DELETED OIL
HOLES AND GROOVES-DES.12523
STANDARD DOWELS - DES.13775.

STRENGTHENED
THICKENED BEARING PANEL
WEBS AND LOWER JOINT
FACE   DES.13796 & DES.14111

LT1 LEAD PLATED
DES.13964.

No.4 MAIN BEARING.
INCREASED RADIUS AT
BACK OF THRUST FACE.
DES.14235.

WHEELCASE JOINT.
3/8" PARALLEL STUDS.
DES.12170.

DEEPER BOSSES AND LONGER MAIN BEARING CAP
STUDS AND EXHAUST SIDE   CYLINDER STUDS
IN Nos 2, 3, 5 & 6 PANELS (SIMILAR TO STD No.4)
                          DES. 14007.

GALLERY PIPE, AND MAIN BEARING
OILWAY DRILLING DELETED.
DES. 14421.

STRENGTHENING, & END TO END OIL FEED FEATURES

CRANKCASE & BEARINGS   1st. 24 R.M.14.S.M. ENGINES.

OCT. 1943
DRG. REF. Lov/Dnl 356.

SHEET REF RM 14 SM /4000/2

30

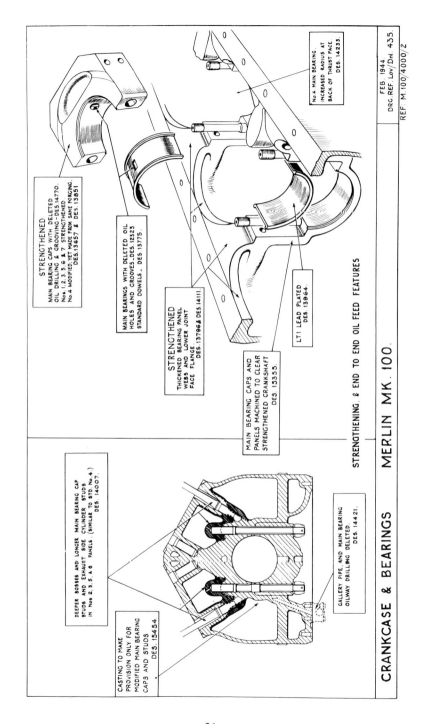

STRENGTHENED
MAIN BEARING CAPS WITH DELETED OIL DRILLING & GROOVING - DES.14770.
Nos. 1,2,3,5,6 & 7 STRENGTHENED
No.4 MODIFIED, YET MADE FROM SAME FORGING.
DES.13657 & DES.13851

MAIN BEARINGS WITH DELETED OIL HOLES AND GROOVES. DES.12323
STANDARD DOWELS - DES.13775.

STRENGTHENED
THICKENED BEARING PANEL WEBS AND LOWER JOINT FACE FLANGE.
DES.13796 & DES.14111.

No.4 MAIN BEARING. INCREASED RADIUS AT BACK OF THRUST FACE.
DES.14233.

LT1 LEAD PLATED
DES.13964.

MAIN BEARING CAPS AND PANELS MACHINED TO CLEAR STRENGTHENED CRANKSHAFT
DES.13355.

STRENGTHENING & END TO END OIL FEED FEATURES

DEEPER BOSSES AND LONGER MAIN BEARING CAP STUDS AND EXHAUST SIDE CYLINDER STUDS IN Nos 2,3,5 & 6 PANELS. (SIMILAR TO STD. No.4.)
DES.14007.

CASTING TO MAKE PROVISION ONLY FOR MODIFIED MAIN BEARING CAPS AND STUDS
DES.15454.

GALLERY PIPE, 4ND MAIN BEARING OILWAY DRILLING DELETED
DES.14421.

CRANKCASE & BEARINGS   MERLIN MK. 100.

FEB. 1944
DRG. REF. Lov/Dml. 435.

REF. M.100/4000/2

31

STRENGTHENED
MAIN BEARING CAPS WITH DELETED
OIL DRILLING & GROOVING-DES.14770.
No.1,2,3,5,6 & 7 STRENGTHENED.
No.4 MODIFIED, YET MADE FROM SAME FORGING.
DES.13657 & DES.13851

MAIN BEARINGS WITH DELETED OIL
HOLES AND GROOVES DES.14470
UNDERCUTS TO BACK OF SHELL
DES.14983.
STANDARD DOWELS ~ DES.13775.

STRENGTHENED
THICKENED BEARING PANEL
WEBS AND LOWER JOINT
FACE FLANGE.
DES.13796 & DES.14111.

No.4 MAIN BEARING.
INCREASED RADIUS AT
BACK OF THRUST FACE.
DES. 14233.

MAIN BEARING CAPS AND
PANELS MACHINED TO CLEAR
STRENGTHENED CRANKSHAFT
DES. 15355.

LT1 LEAD PLATED
DES. 13964.

DEEPER BOSSES AND LONGER MAIN BEARING CAP
STUDS AND EXHAUST SIDE CYLINDER STUDS
IN Nos 2,3,5 & 6 PANELS. (SIMILAR TO STD No 4.)
DES. 14007.

CASTING TO MAKE
PROVISION ONLY FOR
MODIFIED MAIN BEARING
CAPS AND STUDS
DES. 15454

GALLERY PIPE AND MAIN BEARING
OILWAY DRILLING DELETED.
DES. 14421.

STRENGTHENING, & END TO END OIL FEED FEATURES

NOV. 1944

DRG. REF. LOW/DRL/691

MERLIN/4000/5

MERLIN MK 100 SERIES

CRANKCASE & BEARINGS

32

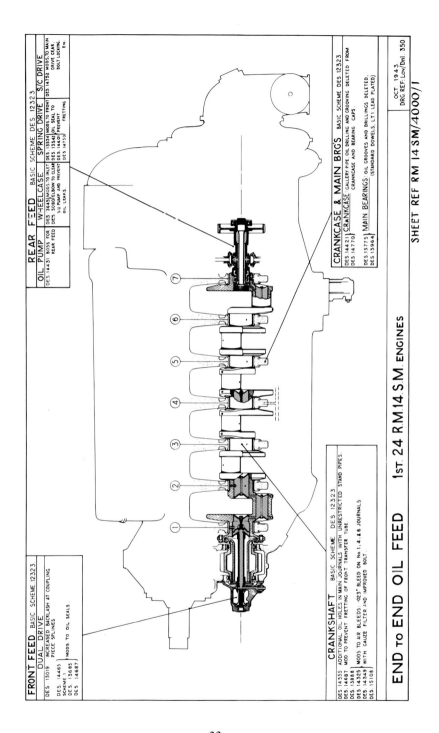

FRONT FEED BASIC SCHEME 12323
DUAL DRIVE
DES. 15019 INCREASED BACKLASH AT COUPLING PIECE SPLINES
DES. 14465 SCHEME 1
DES. 13685 } MODS TO OIL SEALS
DES. 14687

REAR FEED BASIC SCHEME DES 12323

OIL PUMP	WHEELCASE	SPRING DRIVE	S/C DRIVE
DES. 14431 BOSS FOR REAR FEED	DES 3443 MODS TO INLET DES 5 5090 ELBOW TO CLEAR 5.U PUMP AND PREVENT OIL LEAKS	DES 15534 MODS TO FRONT DES. 13542 OIL SEAL TO DES. 14401 PREVENT DES 14759 FRETTING	DES 14792 MODS TO MAIN DRIVE GEAR. BOLT LOCKING. Etc.

CRANKCASE & MAIN BRG'S BASIC SCHEME DES 12323.
CRANKCASE GALLERY PIPE OIL DRILLING AND GROOVING DELETED FROM CRANKCASE AND BEARING CAPS
DES 14421
DES 14770
DES 13775 MAIN BEARINGS OIL GROOVES AND DRILLINGS DELETED.
DES 13964 (STANDARD DOWELS, LT.1 LEAD PLATED)

CRANKSHAFT BASIC SCHEME DES 12323.
DES 14555 ADDITIONAL OIL HOLES IN MAIN JOURNALS WITH UNRESTRICTED STAND PIPES
DES 14687 MOD TO PREVENT FRETTING OF FRONT TRANSFER TUBE
DES 13888 MODS TO AIR BLEEDS. ·023" BLEED ON No 1, 4, & 6 JOURNALS
DES 14325 WITH GAUZE FILTER AND IMPROVED BOLT
DES 14349
DES 15108

END TO END OIL FEED 1st 24 R.M.14.S.M. ENGINES

OCT. 1943
DRG REF: Lov/Dml 350

SHEET REF RM 14 SM/4000/1

(1) (2) (3) (4) (5) (6) (7)

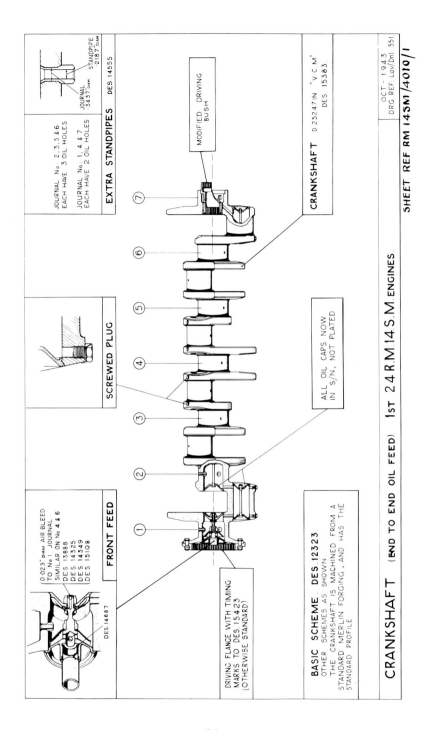

STANDPIPE -2187 DIAM

JOURNAL -3437 DIAM   DES.14555

JOURNAL No: 2, 3, 5 & 6
EACH HAVE 3 OIL HOLES

JOURNAL No: 1, 4 & 7
EACH HAVE 2 OIL HOLES

EXTRA STANDPIPES

MODIFIED DRIVING BUSH

CRANKSHAFT   D.25247 IN "V.C.M"   DES.15383

SCREWED PLUG

ALL OIL CAPS NOW IN S/N, NOT PLATED

FRONT FEED

0.023" DIAM AIR BLEED
TO No 1 JOURNAL
SIMILAR ON No 4 & 6
DES.13888
DES.14325
DES.14349
DES.15108

DES.14687

DRIVING FLANGE WITH TIMING
MARKS TO DES.15423
(OTHERWISE STANDARD)

BASIC SCHEME. DES.12323
OTHER SCHEMES AS SHOWN.
THE CRANKSHAFT IS MACHINED FROM A
STANDARD MERLIN FORGING, AND HAS THE
STANDARD PROFILE

CRANKSHAFT (END TO END OIL FEED) 1ST 24 R.M.14 S.M. ENGINES

OCT - 1943
DRG REF Lov/Dnl 351

SHEET REF RM 14SM/4010/1

34

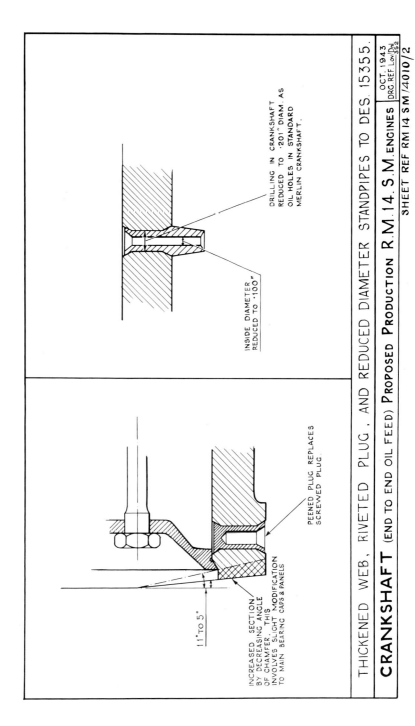

DRILLING IN CRANKSHAFT
REDUCED TO ·201″ DIAM. AS
OIL HOLES IN STANDARD
MERLIN CRANKSHAFT.

INSIDE DIAMETER
REDUCED TO ·100″

PEENED PLUG REPLACES
SCREWED PLUG.

INCREASED SECTION
BY DECREASING ANGLE
OF CHAMFER. THIS
INVOLVES SLIGHT MODIFICATION
TO MAIN BEARING CAPS & PANELS.

11° TO 5°

THICKENED WEB, RIVETED PLUG., AND REDUCED DIAMETER STANDPIPES TO DES. 15355.

CRANKSHAFT (END TO END OIL FEED) **PROPOSED PRODUCTION R.M.14.S.M.ENGINES**

SHEET REF RM 14 S M /4010/2

OCT. 1943
DRG.REF Lov/Dhl 352

35

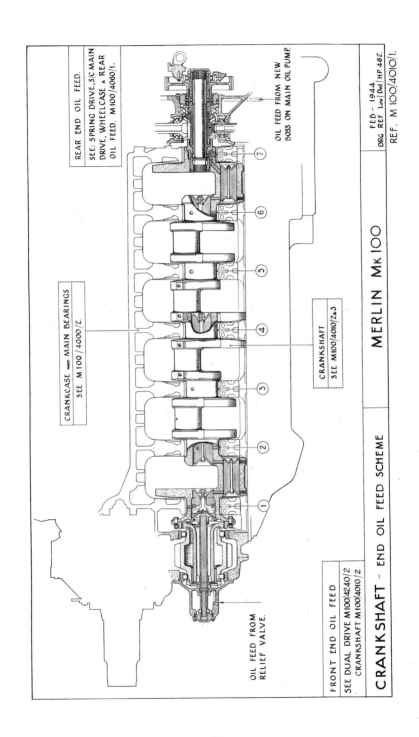

REAR END OIL FEED.

SEE: SPRING DRIVE, S/C MAIN DRIVE, WHEELCASE & REAR OIL FEED. M 100/4060/1.

OIL FEED FROM NEW BOSS ON MAIN OIL PUMP.

FEB - 1944
DRG REF Low/Dal/HP 462.

REF. M 100/4010/1.

MERLIN Mk 100

CRANKCASE AND MAIN BEARINGS
SEE M 100 / 4000 / 2.

CRANKSHAFT
SEE M100/4010/2 & 3

OIL FEED FROM RELIEF VALVE.

FRONT END OIL FEED
SEE DUAL DRIVE M100/4240/2
CRANKSHAFT M100/4010/2

CRANKSHAFT - END OIL FEED SCHEME

36

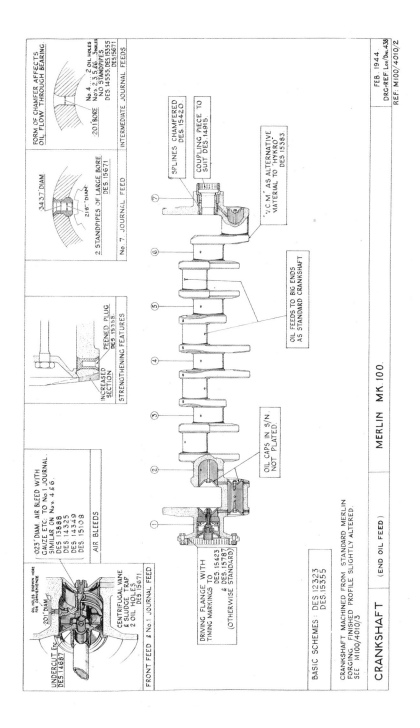

FORM OF CHAMFER AFFECTS
OIL FLOW THROUGH BEARING.

·201 BORE

No. 4 ... 2 OIL HOLES
No's 2, 3, 5 & 6 ... 3 HOLES
NO STANDPIPES
DES. 14555. DES. 15555
DES. 15671

INTERMEDIATE JOURNAL FEEDS

·3437 DIAM

·218" DIAM

2 STANDPIPES OF LARGE BORE
DES. 15671

No. 7 JOURNAL FEED

SPLINES CHAMFERED
DES. 15420

COUPLING PIECE TO
SUIT DES. 14915

'V.C.M.' AS ALTERNATIVE
MATERIAL TO 'HYKRO'
DES. 15383

⑦ ⑥ ⑤ ④ ③ ② ①

OIL FEEDS TO BIG ENDS
AS STANDARD CRANKSHAFT

OIL CAPS IN S/N
NOT PLATED.

PEENED PLUG
DES. 15355

INCREASED
SECTION

STRENGTHENING FEATURES

·023" DIAM. AIR BLEED WITH
GAUZE ETC: TO No. 1 JOURNAL.
SIMILAR ON No's 4 & 6.
DES. 13888
DES. 14325
DES. 14349
DES. 15108

AIR BLEEDS

OIL HOLES SHOWN HERE
FOR CONVENIENCE

·201 DIAM

UNDERCUT Etc.
DES. 14687

CENTRIFUGAL VANE
& SLUDGE TRAP
2 OIL HOLES
DES. 15671

FRONT FEED & No. 1 JOURNAL FEED

DRIVING FLANGE WITH
TIMING MARKINGS TO
DES. 15423
& DES. 15787
(OTHERWISE STANDARD)

BASIC SCHEMES: DES. 12323
DES. 15355

CRANKSHAFT MACHINED FROM STANDARD MERLIN
FORGING, FINISHED PROFILE SLIGHTLY ALTERED.
SEE M100/4010/3

CRANKSHAFT (END OIL FEED)

MERLIN MK. 100.

FEB. 1944.
DRG·REF Lov/Dw.458

REF: M100/4010/2

37

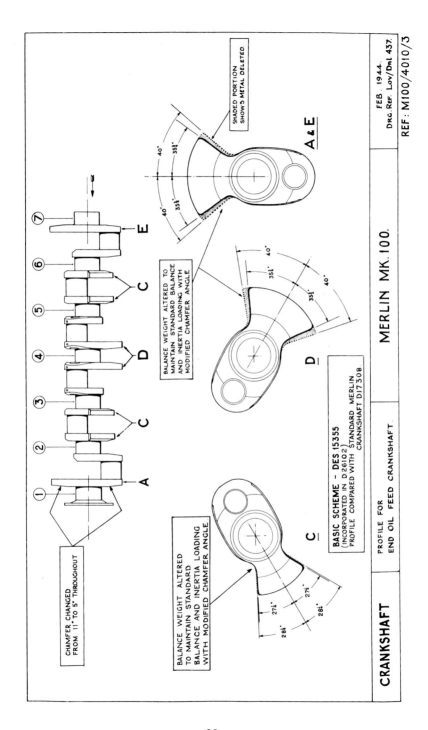

CHAMFER CHANGED
FROM 11° TO 5° THROUGHOUT

BALANCE WEIGHT ALTERED
TO MAINTAIN STANDARD
BALANCE AND INERTIA LOADING
WITH MODIFIED CHAMFER ANGLE

BALANCE WEIGHT ALTERED TO
MAINTAIN STANDARD BALANCE
AND INERTIA LOADING WITH
MODIFIED CHAMFER ANGLE

SHADED PORTION
SHOWS METAL DELETED

A & E

D

C

BASIC SCHEME - DES 15355
(INCORPORATED IN D 26102)
PROFILE COMPARED WITH STANDARD MERLIN
CRANKSHAFT D17308

FEB. 1944.
DRG. REF. LOV/DN1 437.

MERLIN MK. 100.

PROFILE FOR
END OIL FEED CRANKSHAFT

CRANKSHAFT

REF: M100/4010/3

38

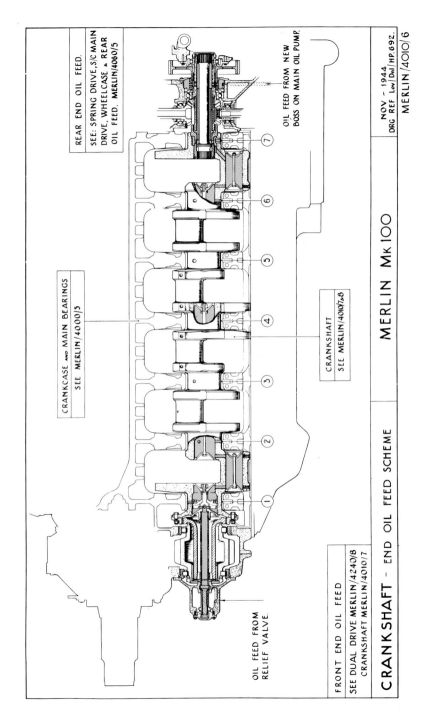

REAR END OIL FEED.

SEE: SPRING DRIVE, S/C MAIN DRIVE, WHEELCASE & REAR OIL FEED. MERLIN/4060/5

OIL FEED FROM NEW BOSS ON MAIN OIL PUMP.

CRANKCASE AND MAIN BEARINGS SEE MERLIN/4000/5

CRANKSHAFT SEE MERLIN/4010/7&8

OIL FEED FROM RELIEF VALVE.

FRONT END OIL FEED

SEE DUAL DRIVE MERLIN/4240/8 CRANKSHAFT MERLIN/4010/7

CRANKSHAFT - END OIL FEED SCHEME

MERLIN Mk 100

NOV - 1944 DRG REF LOW/DML/HP.692.

MERLIN/4010/6

39

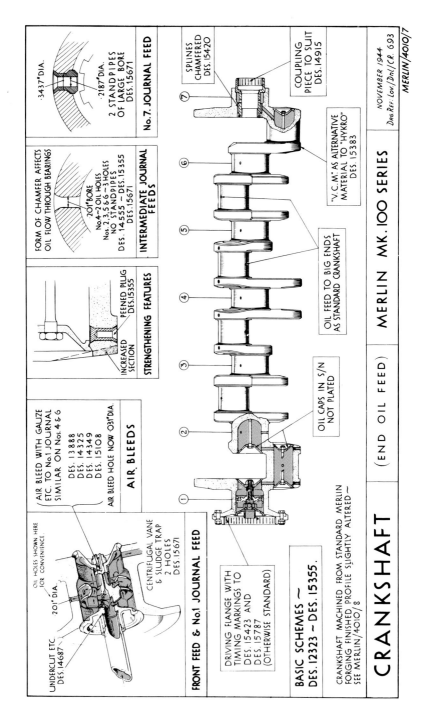

No.7. JOURNAL FEED

·3437" DIA.

·2187" DIA.

2 STANDPIPES OF LARGE BORE DES.15671

⑦ SPLINES CHAMFERED DES.15420

COUPLING PIECE TO SUIT DES.14915

INTERMEDIATE JOURNAL FEEDS

FORM OF CHAMFER AFFECTS OIL FLOW THROUGH BEARINGS

·201" BORE
No.4—2 OIL HOLES
Nos. 2,3,5 & 6 —3 HOLES
NO STANDPIPES
DES. 14555 — DES.15355
DES.15671

"V.C.M." AS ALTERNATIVE MATERIAL TO "HYKRO" DES. 15383

STRENGTHENING FEATURES

PEENED PLUG DES.15355

INCREASED SECTION

OIL FEED TO BIG ENDS AS STANDARD CRANKSHAFT

OIL CAPS IN S/N NOT PLATED

AIR BLEEDS

AIR BLEED WITH GAUZE ETC. TO No.1 JOURNAL SIMILAR ON Nos. 4 & 6

DES. 13888
DES. 14325
DES. 14349
DES. 15108

AIR BLEED HOLE NOW ·031" DIA.

FRONT FEED & No.1 JOURNAL FEED

OIL HOLES SHOWN HERE FOR CONVENIENCE.

·201" DIA.

UNDERCUT ETC. DES.14687

CENTRIFUGAL VANE & SLUDGE TRAP 2 HOLES DES.15671

DRIVING FLANGE WITH TIMING MARKINGS TO DES. 15423 AND DES. 15787 (OTHERWISE STANDARD)

BASIC SCHEMES ~ DES. 12323 ~ DES. 15355.

CRANKSHAFT MACHINED FROM STANDARD MERLIN FORGING FINISHED PROFILE SLIGHTLY ALTERED ~ SEE MERLIN/4010/8

CRANKSHAFT (END OIL FEED) MERLIN MK.100 SERIES

NOVEMBER 1944
Drg.Ref. Lov./Dnl/CR. 693

MERLIN/4010/7

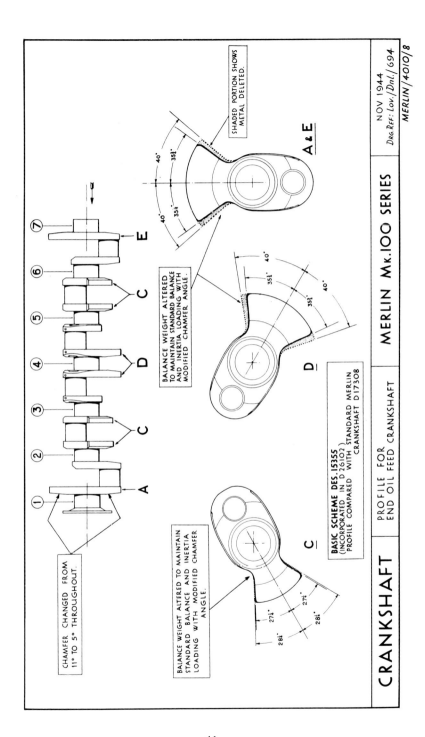

CHAMFER CHANGED FROM 11° TO 5° THROUGHOUT.

BALANCE WEIGHT ALTERED TO MAINTAIN STANDARD BALANCE AND INERTIA LOADING WITH MODIFIED CHAMFER ANGLE.

BALANCE WEIGHT ALTERED TO MAINTAIN STANDARD BALANCE AND INERTIA LOADING WITH MODIFIED CHAMFER ANGLE.

SHADED PORTION SHOWS METAL DELETED.

A & E

D

C

BASIC SCHEME DES. 15355
(INCORPORATED IN D 26102)
PROFILE COMPARED WITH STANDARD MERLIN
CRANKSHAFT D17308

CRANKSHAFT

PROFILE FOR END OIL FEED CRANKSHAFT

MERLIN Mk.100 SERIES

NOV 1944
Dʀɢ.Ref: Lov./Dnl./694

MERLIN/4010/8

41

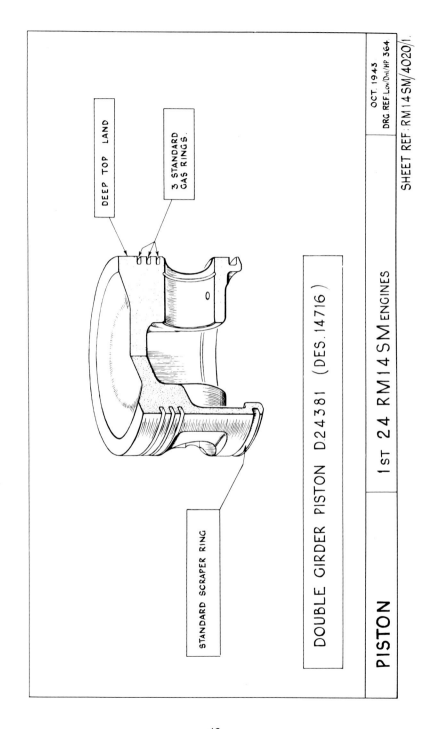

DEEP TOP LAND

3 STANDARD GAS RINGS.

STANDARD SCRAPER RING

DOUBLE GIRDER PISTON D24381 (DES.14716)

PISTON

1ST 24 RM14SM ENGINES

OCT. 1943
DRG REF. Low/Dnl/HP 364

SHEET REF: RM14SM/4020/1.

42

DEEP TOP LAND

3 STANDARD GAS RINGS.

STANDARD SCRAPER RING

16 OIL DRAIN HOLES

RELIEVED IN VICINITY OF GUDGEON PIN BOSSES SIMILAR TO MERLIN XX PISTONS

DOUBLE GIRDER PISTON D26064

PISTON

ALL MERLINS (INCLUDING MERLIN Mᴇ. 100 SERIES)

APRIL 1944
DRG. REF. Lᴏᴡ/Dᴍʟ/HP 532

REF: MERLIN/4020/2

43

DEEP TOP LAND

3 STANDARD GAS RINGS. RS 9025 DTD 485 ·040" to ·045" gap.

SCRAPER RING. RS 9026 DTD 413 ·040" to ·045" gap.

16 OIL DRAIN HOLES

RELIEVED IN VICINITY OF GUDGEON PIN BOSSES. SIMILAR TO MERLIN XX PISTONS.

DOUBLE GIRDER PISTONS D.26064

PISTON

MERLIN MK 100 SERIES

NOV. 1944
Drg. Ref: Low/Dnl/C.R. 695
MERLIN/4020/3

Rolls-Royce Merlin Engine Development for RM14SM Rating.

CONNECTING RODS.   All RM.14SM Engines.

No illustration accompanies this sheet.

Small End Bushes. All RM.14SM Engines will be fitted with L.T.10 bushes to Dwg.14723; this is an improved material which prevents the cracking from the oil holes which has occurred with standard bushes at high power.

Connecting Rods. The First 24 RM14SM Engines will be fitted with standard Merlin 61 type strengthened rods. Later engines will be fitted with similar rods but with an improved form of bolt lug which reduces the stress concentration and is easier to polish in the lug fillet; these rods are just coming on to production for Merlin engines fitted with two piece cylinders and are shown on the latest Merlin 66 General Arrangement Drawing No. D.23881.

Big End and Blade Rod Bearings. All RM14SM Engines are intended to be fitted with the present standard big end assembly.

Further Development work is in hand aimed at improving the lubrication of the blade rod bearing to reduce wear; further details later.

TYPE TEST POSITION, 6.11.43: Extensive development running, including running at RM14SM powers, on the above features.

RM14SM PRODUCTION SPECIFICATION POSITION:

These features are intended for production, but possibly with some modification to blade rod bearing lubrication arrangements, subject to satisfactory Type Test.

CONNECTING RODS   ALL   R.M.14 S.M. ENGINES   Nov. 1943.

45

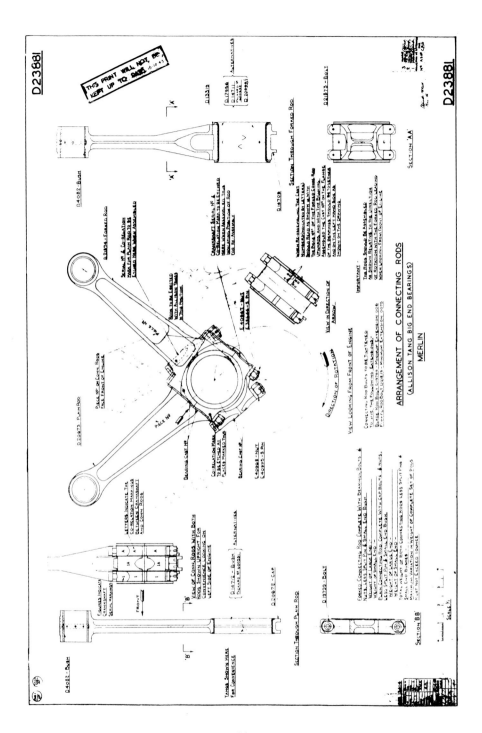

ARRANGEMENT OF CONNECTING RODS
(ALLISON TANG BIG END BEARINGS)
MERLIN

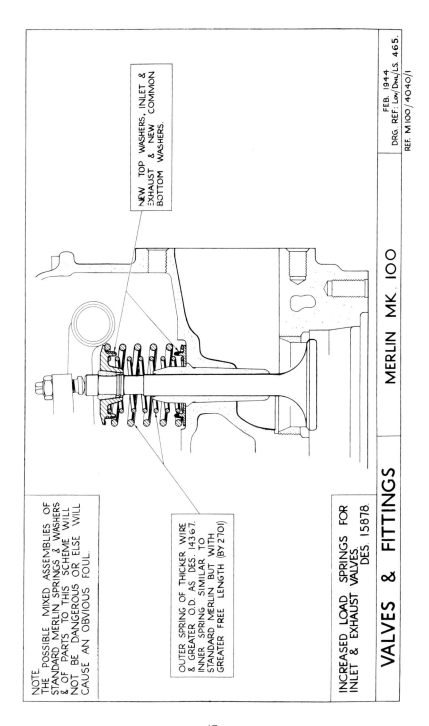

NOTE.
THE POSSIBLE MIXED ASSEMBLIES OF
STANDARD MERLIN SPRINGS & WASHERS
& OF PARTS TO THIS SCHEME WILL
NOT BE DANGEROUS OR ELSE WILL
CAUSE AN OBVIOUS FOUL.

NEW TOP WASHERS. INLET &
EXHAUST & NEW COMMON
BOTTOM WASHERS.

OUTER SPRING OF THICKER WIRE
& GREATER O.D. AS DES. 14367.
INNER SPRING SIMILAR TO
STANDARD MERLIN BUT WITH
GREATER FREE LENGTH (BY 2701)

INCREASED LOAD SPRINGS FOR
INLET & EXHAUST VALVES.
DES. 15878.

VALVES & FITTINGS

MERLIN MK. 100

FEB. 1944
DRG. REF: Lov/DNL/LS. 465.
REF. M100 / 4040/1

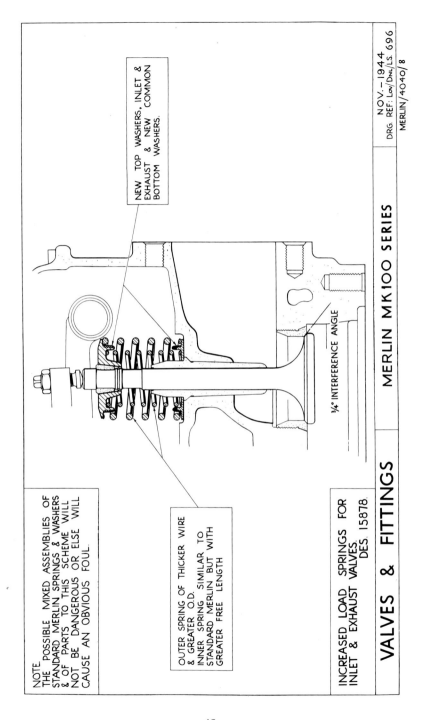

NEW TOP WASHERS, INLET & EXHAUST & NEW COMMON BOTTOM WASHERS.

NOTE.
THE POSSIBLE MIXED ASSEMBLIES OF STANDARD MERLIN SPRINGS & WASHERS & OF PARTS TO THIS SCHEME WILL NOT BE DANGEROUS OR ELSE WILL CAUSE AN OBVIOUS FOUL.

OUTER SPRING OF THICKER WIRE & GREATER O.D. INNER SPRING SIMILAR TO STANDARD MERLIN BUT WITH GREATER FREE LENGTH

¾° INTERFERENCE ANGLE

INCREASED LOAD SPRINGS FOR INLET & EXHAUST VALVES. DES. 15878.

VALVES & FITTINGS

MERLIN MK100 SERIES

NOV.-1944
DRG. REF: Lɑ/Dɴ/LS. 696

MERLIN/4040/8

48

INDUCTION PIPE SEALING PAD. DES. 14.184.

No5. 3 & 4 CYLINDER CORE PLUGS DELETED FROM TOP PLATFORM OF HEAD. DES. 14327.

℄ No. 3 CYLINDER

℄ No. 2 CYLINDER

GROOVE IN CYLINDER HEAD NEAR JOINT FLANGE OF LINER DES. 13427.

5 B.A. PEG.

PEGGED INSERTS IN CYLINDER HEAD FOR EXHAUST MANIFOLD STUDS. REPAIR. Sc. 72.

CYLINDERS 1st 24 R.M.14. S.M. ENGINES

OCT. 1943.
DRG. REF. Lov/Dml. 355.

SHEET REF RM 14 SM /4050/1

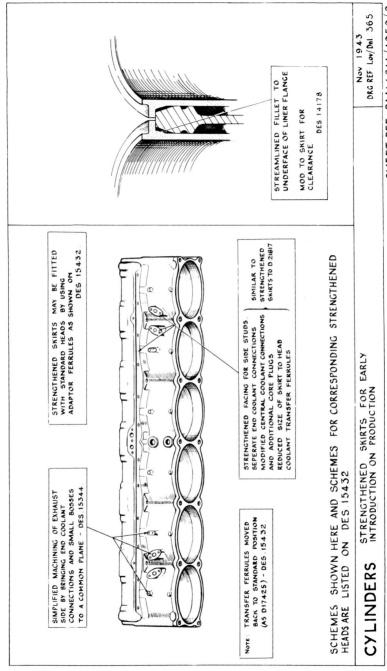

SIMPLIFIED MACHINING OF EXHAUST SIDE BY BRINGING END COOLANT CONNECTIONS AND SMALL BOSSES TO A COMMON PLANE - DES 15344

STRENGTHENED SKIRTS MAY BE FITTED WITH STANDARD HEADS BY USING ADAPTOR FERRULES AS SHOWN ON DES 15432

NOTE: TRANSFER FERRULES MOVED BACK TO STANDARD POSITION (AS D17425) - DES 15432

STRENGTHENED FACING FOR SIDE STUDS
SEPERATE END COOLANT CONNECTIONS
MODIFIED CENTRAL COOLANT CONNECTIONS
AND ADDITIONAL CORE PLUGS
REDUCED SIZE OF SKIRT TO HEAD
COOLANT TRANSFER FERRULES

SIMILAR TO STRENGTHENED SKIRTS TO D 21817

STREAMLINED FILLET TO UNDERFACE OF LINER FLANGE

MOD TO SKIRT FOR CLEARANCE
DES 14178

SCHEMES SHOWN HERE AND SCHEMES FOR CORRESPONDING STRENGTHENED HEADS ARE LISTED ON DES 15432

CYLINDERS STRENGTHENED SKIRTS FOR EARLY INTRODUCTION ON PRODUCTION

Nov 1943
DRG REF Lov/Dml 365

SHEET REF: RM14SM/4050/2.

50

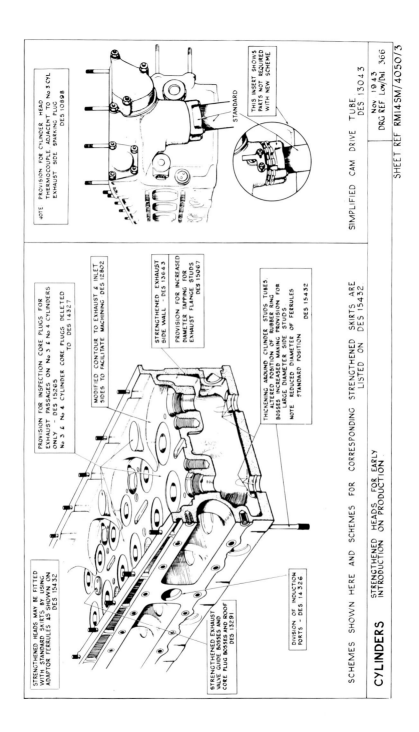

NOTE PROVISION FOR CYLINDER HEAD THERMOCOUPLE ADJACENT TO No 3 CYL EXHAUST SIDE SPARKING PLUG DES.10898

STANDARD

THIS INSERT SHOWS PARTS NOT REQUIRED WITH NEW SCHEME

SIMPLIFIED CAM DRIVE TUBE DES 13043

Nov 1943
DRG REF Low/Dni 366

PROVISION FOR INSPECTION CORE PLUGS FOR EXHAUST PASSAGES ON No 3 £ No 4 CYLINDERS ONLY - DES 15265
No 3 £ No 4 CYLINDER CORE PLUGS DELETED TO DES 14327

STRENGTHENED EXHAUST SIDE WALL - DES 13643

MODIFIED CONTOUR TO EXHAUST & INLET SIDES TO FACILITATE MACHINING DES 12802

PROVISION FOR INCREASED DIAMETER TAPPING FOR EXHAUST FLANGE STUDS DES 15067

THICKENING AROUND CYLINDER STUDS TUBES ALTERED POSITION OF RUBBER RING BOSSES INCREASED MAKING PROVISION FOR LARGE DIAMETER SIDE STUDS
NOTE REDUCED DIAMETER OF FERRULES STANDARD POSITION DES 15432

STRENGTHENED HEADS MAY BE FITTED WITH STANDARD SKIRTS BY USING ADAPTOR FERRULES AS SHOWN ON DES 15432

STRENGTHENED EXHAUST VALVE GUIDE BOSSES AND CORE PLUG BOSSES AND ROOF DES 13291

DIVISION OF INDUCTION PORTS - DES 14326

SCHEMES SHOWN HERE AND SCHEMES FOR CORRESPONDING STRENGTHENED SKIRTS ARE LISTED ON DES 15432

CYLINDERS

STRENGTHENED HEADS FOR EARLY INTRODUCTION ON PRODUCTION

SHEET REF RM14.SM/4050/3

51

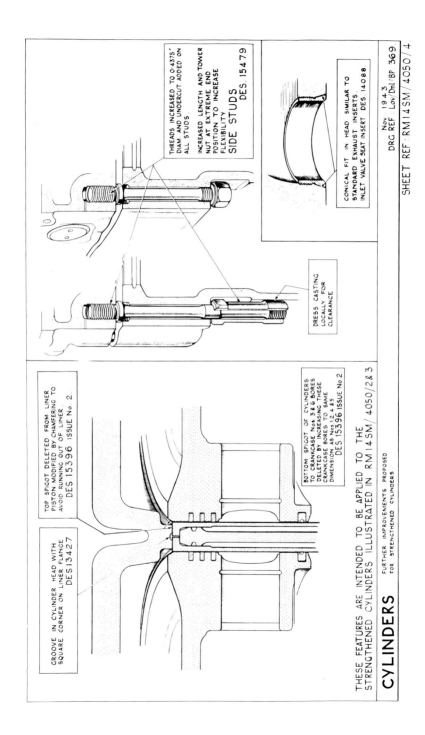

THREADS INCREASED TO 0.4375"
DIAM. AND UNDERCUT ADDED ON
ALL STUDS.

INCREASED LENGTH AND TOWER
NUT AT EXTREME END
POSITION TO INCREASE
FLEXIBILITY.

SIDE STUDS
DES. 15479

CONICAL FIT IN HEAD SIMILAR TO
STANDARD EXHAUST INSERTS.
INLET VALVE SEAT INSERT DES. 14088

Nov. 1943
DRG. REF. Lov/Dnl/BP 369

DRESS CASTING
LOCALLY FOR
CLEARANCE

TOP SPIGOT DELETED FROM LINER.
PISTON MODIFIED BY CHAMFERING TO
AVOID RUNNING OUT OF LINER
DES. 15396 ISSUE No. 2.

BOTTOM SPIGOT OF CYLINDERS
TO CRANKCASE Nos. 3 & 6 BORES
DELETED BY INCREASING THESE
CRANKCASE BORES TO SAME
DIMENSIONS AS Nos. 1,2,4 & 5
DES. 15396 ISSUE No. 2

GROOVE IN CYLINDER HEAD WITH
SQUARE CORNER ON LINER FLANGE
DES. 13427

THESE FEATURES ARE INTENDED TO BE APPLIED TO THE
STRENGTHENED CYLINDERS ILLUSTRATED IN RM 14 SM/4050/2 & 3

CYLINDERS   FURTHER IMPROVEMENTS PROPOSED
             FOR STRENGTHENED CYLINDERS

SHEET REF: RM 14 SM/4050/4.

52

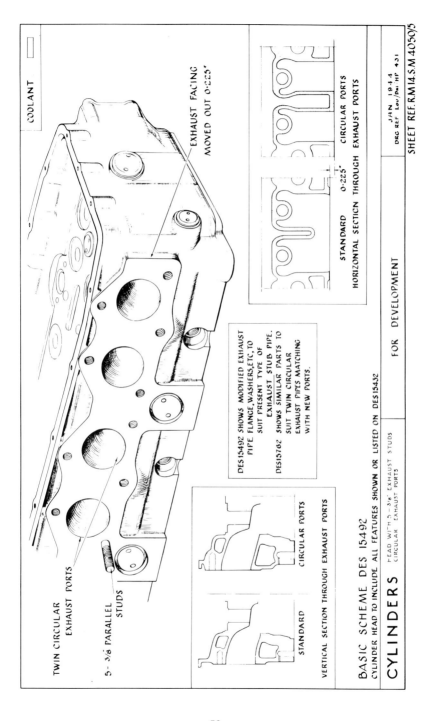

COOLANT

EXHAUST FACING MOVED OUT 0·225"

TWIN CIRCULAR EXHAUST PORTS

5 - 3/8 PARALLEL STUDS

CIRCULAR PORTS

DES 15492 SHOWS MODIFIED EXHAUST PIPE FLANGE, WASHERS, ETC, TO SUIT PRESENT TYPE OF EXHAUST STUB PIPE.

DES 15762 SHOWS SIMILAR PARTS TO SUIT TWIN CIRCULAR EXHAUST PIPES MATCHING WITH NEW PORTS.

STANDARD

CIRCULAR PORTS

VERTICAL SECTION THROUGH EXHAUST PORTS

STANDARD   0·225"   CIRCULAR PORTS

HORIZONTAL SECTION THROUGH EXHAUST PORTS

BASIC SCHEME DES 15492
CYLINDER HEAD TO INCLUDE ALL FEATURES SHOWN OR LISTED ON DES 15432

CYLINDERS   HEAD WITH 5 - 3/8" EXHAUST STUDS
CIRCULAR EXHAUST PORTS

FOR DEVELOPMENT

JAN 1944
DRG. REF. Lav/Dal HF 431

SHEET REF. R.M 14.S.M 4050/5

53

STREAMLINED FILLET TO
UNDERFACE OF LINER FLANGE
DES 14178

MOD TO SKIRT FOR
CLEARANCE

DES 14178

FEB. 1944.
DRG REF: Lov/Dhl 439.

REF: M.100/4050/1

STRENGTHENED SKIRTS MAY BE FITTED
WITH STANDARD HEADS BY USING
ADAPTOR FERRULES AS SHOWN ON
DES. 15432.

COOLANT CONNS TO - DES. 15456
SIMPLIFIED MACHINING OF EXHAUST
SIDE BY BRINGING END COOLANT
CONNECTIONS AND SMALL BOSSES
TO A COMMON PLANE - DES 15344.

STRENGTHENED FACING FOR SIDE STUDS
SEPERATE END COOLANT CONNECTIONS
MODIFIED CENTRAL COOLANT CONNECTIONS
AND ADDITIONAL CORE PLUGS
REDUCED SIZE OF SKIRT TO HEAD
COOLANT TRANSFER FERRULES
TRANSFER FERRULES IN STANDARD POSITION
DES. 15432

SIMILAR TO
STRENGTHENED
SKIRTS TO D.21817

CYLINDERS   MERLIN MK. 100.

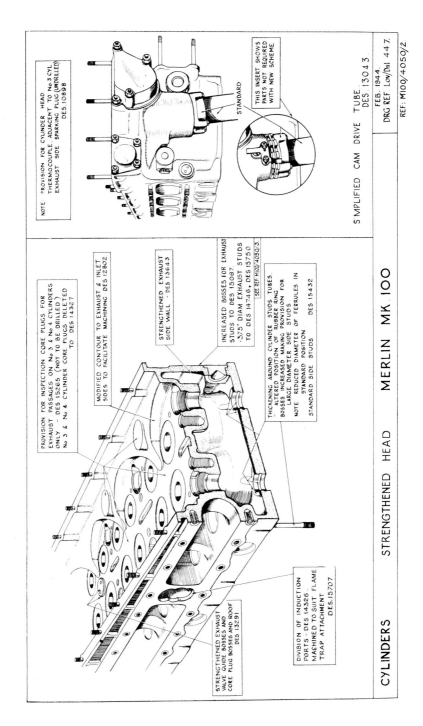

NOTE PROVISION FOR CYLINDER HEAD THERMOCOUPLE ADJACENT TO No.3 CYL. EXHAUST SIDE SPARKING PLUG (UNDRILLED) DES. 10898

STANDARD

THIS INSERT SHOWS PARTS NOT REQUIRED WITH NEW SCHEME

SIMPLIFIED CAM DRIVE TUBE DES. 13043

FEB. 1944.
DRG. REF. Lov/Inl. 447.

REF: M100/4050/2.

PROVISION FOR INSPECTION CORE PLUGS FOR EXHAUST PASSAGES ON No.3 & No.4 CYLINDERS ONLY - DES. 15265 (NOT TO BE DRILLED) No.3 & No.4 CYLINDER CORE PLUGS DELETED TO DES. 14327

MODIFIED CONTOUR TO EXHAUST & INLET SIDES TO FACILITATE MACHINING DES. 12802

STRENGTHENED EXHAUST SIDE WALL - DES. 13643

INCREASED BOSSES FOR EXHAUST STUDS TO DES 15087 .575" DIAM. EXHAUST STUDS TO DES. 14746, DES. 15750
SEE REF. M100/4050/3

THICKENING AROUND CYLINDER STUDS TUBES. ALTERED POSITION OF RUBBER RING BOSSES INCREASED MAKING PROVISION FOR LARGE DIAMETER SIDE STUDS. NOTE REDUCED DIAMETER OF FERRULES IN STANDARD POSITION STANDARD SIDE STUDS DES. 15432.

DIVISION OF INDUCTION PORTS - DES. 14326. MACHINED TO SUIT FLAME TRAP ATTACHMENT DES. 15707

STRENGTHENED EXHAUST VALVE GUIDE BOSSES AND CORE PLUG BOSSES AND ROOF DES. 13291

CYLINDERS   STRENGTHENED   HEAD   MERLIN   MK. 100

55

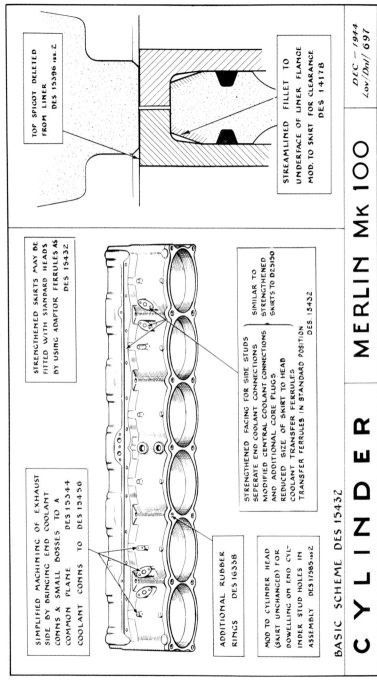

DEC — 1944
Lov/Dnl/ 697

TOP SPIGOT DELETED FROM LINER DES 15396 iss.2

STREAMLINED FILLET TO UNDERFACE OF LINER FLANGE MOD. TO SKIRT FOR CLEARANCE DES 14178

STRENGTHENED SKIRTS MAY BE FITTED WITH STANDARD HEADS BY USING ADAPTOR FERRULES AS DES 15432

SIMILAR TO STRENGTHENED SKIRTS TO D25150

STRENGTHENED FACING FOR SIDE STUDS SEPERATE END COOLANT CONNECTIONS MODIFIED CENTRAL COOLANT CONNECTIONS AND ADDITIONAL CORE PLUGS REDUCED SIZE OF SKIRT TO HEAD COOLANT TRANSFER FERRULES TRANSFER FERRULES IN STANDARD POSITION DES. 15432.

SIMPLIFIED MACHINING OF EXHAUST SIDE BY BRINGING END COOLANT CONNS & SMALL BOSSES TO A COMMON PLANE DES 15344 COOLANT CONNS TO DES 15456

ADDITIONAL RUBBER RINGS DES 16338

MOD TO CYLINDER HEAD (SKIRT UNCHANGED) FOR DOWELLING ON END CYL- INDER STUD HOLES IN ASSEMBLY DES 17585 iss2

BASIC SCHEME DES 15432

CYLINDER MERLIN Mk 100

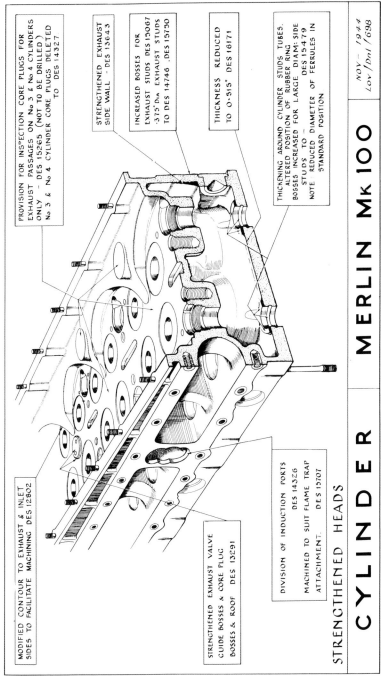

PROVISION FOR INSPECTION CORE PLUGS FOR
EXHAUST PASSAGES ON N₀ 3 & N₀ 4 CYLINDERS
ONLY - DES 15265 (NOT TO BE DRILLED)
N₀ 3 & N₀ 4 CYLINDER CORE PLUGS DELETED
TO DES 14327.

STRENGTHENED EXHAUST
SIDE WALL - DES 13643

INCREASED BOSSES FOR
EXHAUST STUDS DES 13067
·375″D₁₄ EXHAUST STUDS
TO DES 14746 .DES 15730

THICKNESS REDUCED
TO 0·515″ DES 16171

THICKENING AROUND CYLINDER STUDS TUBES.
ALTERED POSITION OF RUBBER RING
BOSSES INCREASED FOR LARGE DIAM: SIDE
STUDS TO - DES 15479
NOTE REDUCED DIAMETER OF FERRULES IN
STANDARD POSITION

MODIFIED CONTOUR TO EXHAUST & INLET
SIDES TO FACILITATE MACHINING DES 12802

STRENGTHENED EXHAUST VALVE
GUIDE BOSSES & CORE PLUG
BOSSES & ROOF DES 13291

DIVISION OF INDUCTION PORTS
DES 14326
MACHINED TO SUIT FLAME TRAP
ATTACHMENT. DES 15707

STRENGTHENED HEADS

CYLINDER | MERLIN Mk 100

NOV - 1944
Lov/Dnl/698

MERLIN/4050/10

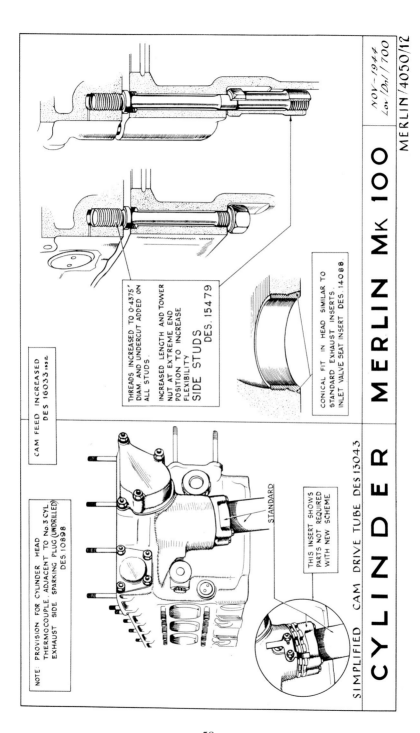

NOTE: PROVISION FOR CYLINDER HEAD THERMOCOUPLE. ADJACENT TO No. 3 CYL. EXHAUST SIDE SPARKING PLUG (UNDRILLED) DES. 10898

CAM FEED INCREASED DES. 16033

THREADS INCREASED TO 0·4375" DIAM. AND UNDERCUT ADDED ON ALL STUDS.

INCREASED LENGTH AND TOWER NUT AT EXTREME END POSITION TO INCREASE FLEXIBILITY

SIDE STUDS DES. 15479

CONICAL FIT IN HEAD SIMILAR TO STANDARD EXHAUST INSERTS. INLET VALVE SEAT INSERT DES. 14088

STANDARD

THIS INSERT SHOWS PARTS NOT REQUIRED WITH NEW SCHEME

SIMPLIFIED CAM DRIVE TUBE DES. 13043

CYLINDER MERLIN MK 100

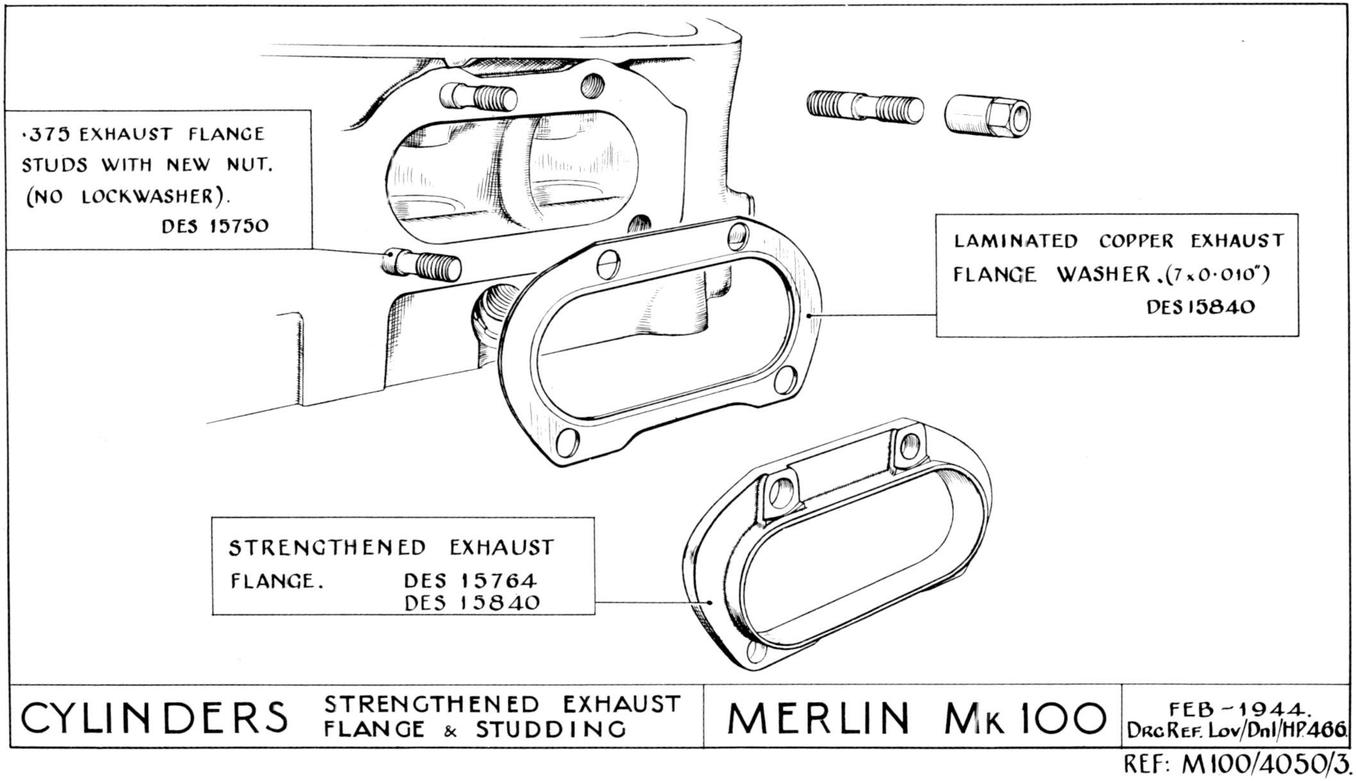

LAMINATED COPPER EXHAUST
FLANGE WASHER. (7 x 0·010")
DES 15840

·375 EXHAUST FLANGE
STUDS WITH NEW NUT.
(NO LOCKWASHER).
DES 15750

STRENGTHENED EXHAUST
FLANGE.   DES 15764
           DES 15840

CYLINDERS   STRENGTHENED EXHAUST
             FLANGE & STUDDING

MERLIN MK 100

FEB-1944.
DRGREF. Lov/Dnl/HP.466

REF: M100/4050/3.

59

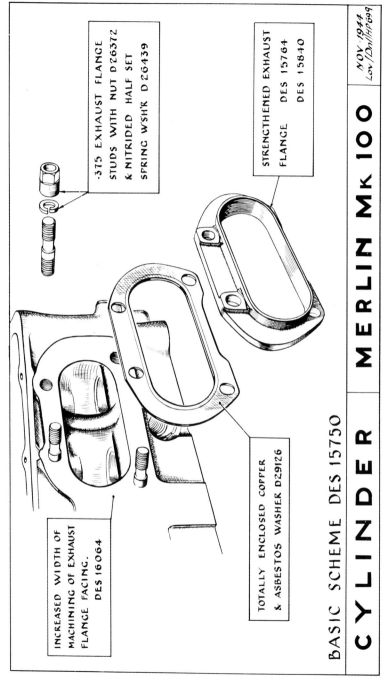

·375 EXHAUST FLANGE STUDS WITH NUT D26372 & NITRIDED HALF SET SPRING W'SHR D26439

STRENGTHENED EXHAUST FLANGE DES 15764 DES 15840

INCREASED WIDTH OF MACHINING OF EXHAUST FLANGE FACING. DES 16064

TOTALLY ENCLOSED COPPER & ASBESTOS WASHER D29126

BASIC SCHEME DES 15750

CYLINDER MERLIN Mk 100

BOTH ROCKER COVER OUTLETS
COUPLED BY AIRCRAFT PIPEWORK
TO COMMON BREATHER PIPE
(TO BE NOT LESS THAN 1½" DIAM.)

CONNECTION DIE CAST
TO DES. 15635

NEW CRANKCASE BREATHER
OUTLET WITH PIPE FEEDING
INTO FRONT END OF BOTH
ROCKER COVERS
(BREATHER BODY DELETED)
DES 15343
DES 15813

INSIDE VIEW OF
OUTLET CONNECTION REAR END
OF ROCKER COVER

INSIDE VIEW OF
INLET CONNECTION FRONT END
OF ROCKER COVER

BASIC SCHEME   DES. 15286

CYLINDERS (ROCKER COVER BREATHERS)   MERLIN MK 100

FEB. 1944
Drg Ref Lov/Dnl/ 456

REF: M100./4050/ 4

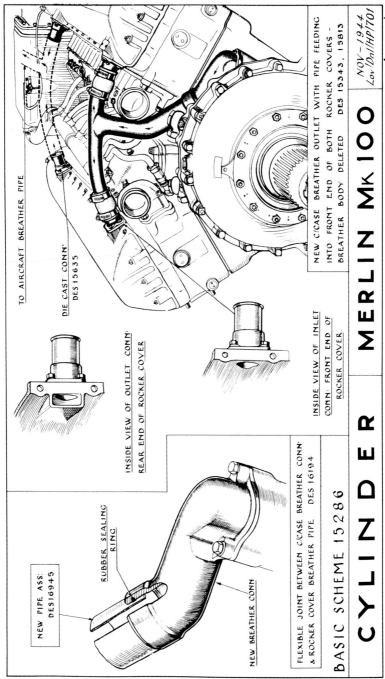

TO AIRCRAFT BREATHER PIPE

DIE CAST CONN:
DES 15635

INSIDE VIEW OF OUTLET CONN:
REAR END OF ROCKER COVER

INSIDE VIEW OF INLET
CONN: FRONT END OF
ROCKER COVER.

NEW C/CASE BREATHER OUTLET WITH PIPE FEEDING
INTO FRONT END OF BOTH ROCKER COVERS -
BREATHER BODY DELETED   DES 15343, 15B13

NEW PIPE ASS:
DES 16945

RUBBER SEALING
RING

NEW BREATHER CONN

FLEXIBLE JOINT BETWEEN C/CASE BREATHER CONN:
& ROCKER COVER BREATHER PIPE   DES 16194

BASIC SCHEME 15286

CYLINDER   MERLIN Mk 100

NOV-1944
Lov/Dn/HP/701

MERLIN/4050/13

STIFFENED PANEL WITH INCREASED
CAPACITY BALL BEARING.
DES.14211.
FLANGED BEARING HOUSING.
DES.14790
MAIN DRIVE GEAR
DES.14752.

SUPERCHARGER MAIN DRIVE.

ELBOW TO CLEAR S.U. INJECTOR PUMP
DES.13645.
IMPROVED BOLTING   DES.15080.

MODIFIED OIL INLET ELBOW

FUEL PUMP DRIVE O·9167:1 RATIO
TO SUIT S.U. INJECTOR PUMP. DES.14179.

0·95" DIAM. SHAFT.   DES.15149.
8½" STOPS   DES.13870.

SPRING DRIVE TORSION SHAFT

MODS TO OIL SEALS { DES.14250.
                    DES.13542.
                    DES.13534.
CASEHARDENING ON END OF (PART)
MAIN DRIVE SHAFT   DES.14401.

OIL SEAL MODS.

3/8" PARALLEL STUDS IN C/CASE TOP HALF
AS DES.12170.
IMPROVED LOCKING TO DES.15098 & DES.15184.

WHEELCASE TO C/CASE JOINT

REAR OIL FEED TO BASIC SCHEME. DES.12323.

SPRING DRIVE , SUPERCHARGER MAIN DRIVE.
WHEELCASE AND REAR OIL FEED

1ST. 24. R.M. 14. S.M. ENGINES

OCT. 1943.

DRG.REF. Lov/Dn1.360

SHEET REF RM 14 SM /4060/1

63

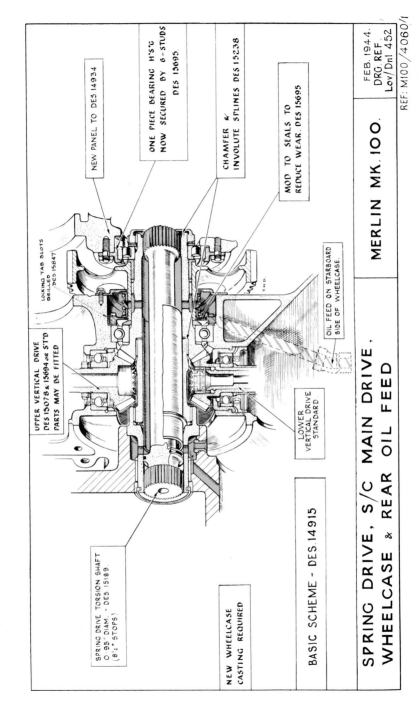

NEW PANEL TO DES 14934

ONE PIECE BEARING H'S'G NOW SECURED BY 6-STUDS DES 15695

CHAMFER & INVOLUTE SPLINES DES 15238

MOD TO SEALS TO REDUCE WEAR. DES 15695

LOCKING TAB SLOTS DRILLED DES 15847

UPPER VERTICAL DRIVE DES 15078 & 15694 or ST'D PARTS MAY BE FITTED

T.H.D.

OIL FEED ON STARBOARD SIDE OF WHEELCASE.

LOWER VERTICAL DRIVE STANDARD

SPRING DRIVE TORSION SHAFT 0.95" DIAM. - DES 15189. (8½" STOPS)

NEW WHEELCASE CASTING REQUIRED

BASIC SCHEME - DES.14915

MERLIN MK. 100.

SPRING DRIVE, S/C MAIN DRIVE, WHEELCASE & REAR OIL FEED

FEB. 1944.
DRG. REF.
Lov/Dnl 452

REF: M100/4060/1

64

NEW PANEL TO DES 14934

ONE PIECE BEARING H'S'G NOW SECURED BY 6-STUDS DES 15695.

CHAMFER & INVOLUTE SPLINES. DES. 15238
IMPROVED LOCATION OF CIRCLIP. DES. 17186

MOD TO SEALS TO REDUCE WEAR. DES 15695

LOCKING TAB SLOTS DRILLED DES 15847

UPPER VERTICAL DRIVE DES 15078 & 15694 or STD PARTS MAY BE FITTED

THD.

OIL FEED ON STARBOARD SIDE OF WHEELCASE.

LOWER VERTICAL DRIVE STANDARD

·900" DIAM. TORSION SHAFT WITH RADIUSED SERRATIONS. DES.16046

DES.16352 MOD. TO EASE PROD.

MODIFIED EXTERNAL SPLINES TO COUPLING SLEEVE TO ALLOW 10°SLOP DES. 16046

NEW WHEELCASE CASTING REQUIRED

BASIC SCHEME – DES.14915

MERLIN MK. 100.

SPRING DRIVE, S/C MAIN DRIVE,
WHEELCASE & REAR OIL FEED.

NOV. 1944
Drg. Ref. Lox/Dnl/702

MERLIN/4060/5

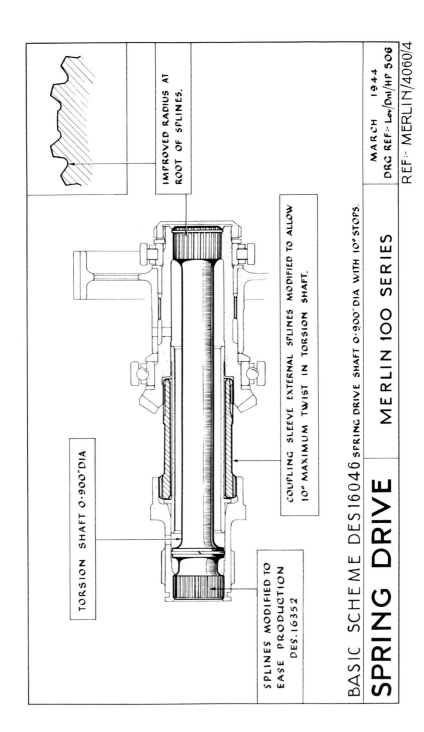

IMPROVED RADIUS AT ROOT OF SPLINES.

TORSION SHAFT 0·900" DIA

SPLINES MODIFIED TO EASE PRODUCTION DES.16352

COUPLING SLEEVE EXTERNAL SPLINES MODIFIED TO ALLOW 10° MAXIMUM TWIST IN TORSION SHAFT.

BASIC SCHEME DES 16046 SPRING DRIVE SHAFT 0·900" DIA WITH 10° STOPS.

SPRING DRIVE MERLIN 100 SERIES

MARCH 1944
DRG REF:- Lov/Dnl/HP 506

REF:- MERLIN/4060/4

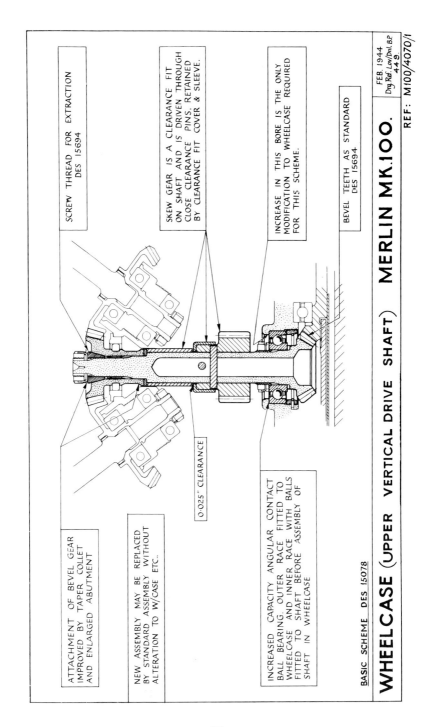

SCREW THREAD FOR EXTRACTION
DES 15694

SKEW GEAR IS A CLEARANCE FIT
ON SHAFT AND IS DRIVEN THROUGH
CLOSE CLEARANCE PINS, RETAINED
BY CLEARANCE FIT COVER & SLEEVE.

INCREASE IN THIS BORE IS THE ONLY
MODIFICATION TO WHEELCASE REQUIRED
FOR THIS SCHEME.

BEVEL TEETH AS STANDARD
DES 15694

ATTACHMENT OF BEVEL GEAR
IMPROVED BY TAPER COLLET
AND ENLARGED ABUTMENT

NEW ASSEMBLY MAY BE REPLACED
BY STANDARD ASSEMBLY WITHOUT
ALTERATION TO W/CASE ETC..

0·025" CLEARANCE

INCREASED CAPACITY ANGULAR CONTACT
BALL BEARING. OUTER RACE FITTED TO
WHEELCASE AND INNER RACE WITH BALLS
FITTED TO SHAFT BEFORE ASSEMBLY OF
SHAFT IN WHEELCASE

BASIC SCHEME DES 15078

WHEELCASE (UPPER VERTICAL DRIVE SHAFT) MERLIN MK.100.

FEB. 1944
Drg. Ref. Low/Dni. B.P.
449

REF: M100/4070/1

67

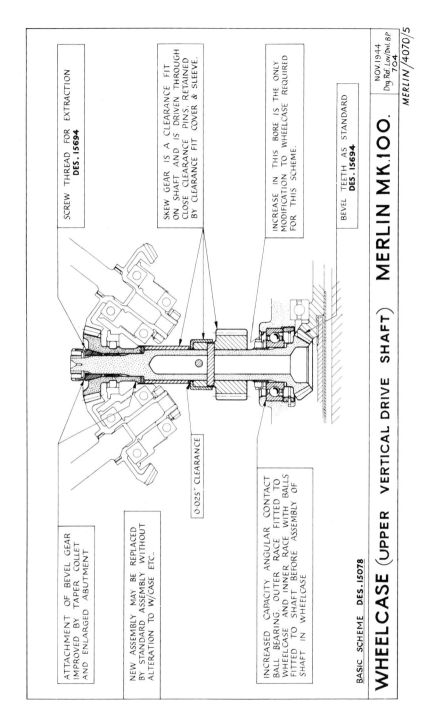

SCREW THREAD FOR EXTRACTION
DES.15694

SKEW GEAR IS A CLEARANCE FIT ON SHAFT AND IS DRIVEN THROUGH CLOSE CLEARANCE PINS, RETAINED BY CLEARANCE FIT COVER & SLEEVE.

INCREASE IN THIS BORE IS THE ONLY MODIFICATION TO WHEELCASE REQUIRED FOR THIS SCHEME.

BEVEL TEETH AS STANDARD
DES.15694

ATTACHMENT OF BEVEL GEAR IMPROVED BY TAPER COLLET AND ENLARGED ABUTMENT

NEW ASSEMBLY MAY BE REPLACED BY STANDARD ASSEMBLY WITHOUT ALTERATION TO W/CASE ETC..

0·025" CLEARANCE

INCREASED CAPACITY ANGULAR CONTACT BALL BEARING. OUTER RACE FITTED TO WHEELCASE AND INNER RACE WITH BALLS FITTED TO SHAFT BEFORE ASSEMBLY OF SHAFT IN WHEELCASE

BASIC SCHEME **DES.15078**

WHEELCASE (UPPER VERTICAL DRIVE SHAFT) MERLIN MK.100.

NOV.1944
Drg Ref Low/Dnl. B.P
704

MERLIN/4070/5

68

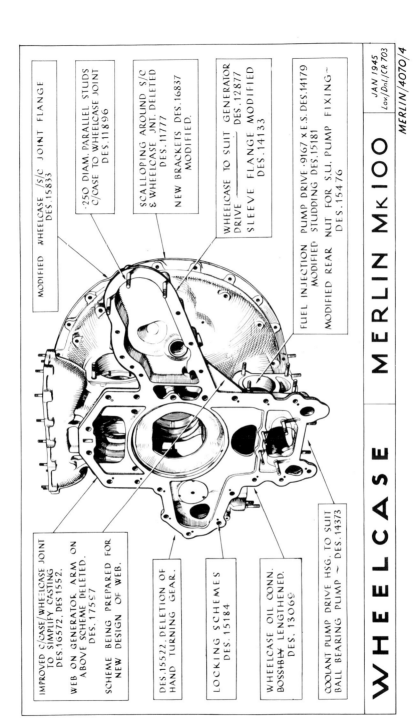

MODIFIED WHEELCASE /S/C JOINT FLANGE DES.15833

.250 DIAM. PARALLEL STUDS C/CASE TO WHEELCASE JOINT DES.11896

SCALLOPING AROUND S/C & WHEELCASE JNT. DELETED DES.11777

NEW BRACKETS DES.16837 MODIFIED.

WHEELCASE TO SUIT GENERATOR DRIVE _____ DES.12877
SLEEVE FLANGE MODIFIED DES.14133

FUEL INJECTION PUMP DRIVE .9167 x E.S. DES.14179 MODIFIED STUDDING DES.15181

MODIFIED REAR NUT FOR S.U. PUMP FIXING— DES.15476

IMPROVED C/CASE/WHEELCASE JOINT TO SIMPLIFY CASTING DES.16572, DES1552.

WEB ON GENERATOR ARM ON ABOVE SCHEME DELETED. DES.17557

SCHEME BEING PREPARED FOR NEW DESIGN OF WEB.

DES.15522. DELETION OF HAND TURNING GEAR.

LOCKING SCHEMES DES. 15184

WHEELCASE OIL CONN. BOSSHEX LENGTHENED. DES. 13069

COOLANT PUMP DRIVE HSG. TO SUIT BALL BEARING PUMP ~ DES.14373

WHEELCASE | MERLIN MkIOO

JAN 1945
Lov/Dn.I/CR 703

MERLIN/4070/4

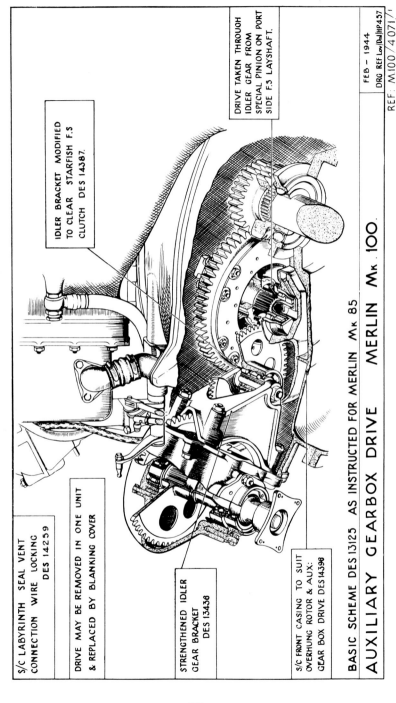

S/C LABYRINTH SEAL VENT
CONNECTION WIRE LOCKING
DES 14259

DRIVE MAY BE REMOVED IN ONE UNIT
& REPLACED BY BLANKING COVER

STRENGTHENED IDLER
GEAR BRACKET
DES 13436

S/C FRONT CASING TO SUIT
OVERHUNG ROTOR & AUX:
GEAR BOX DRIVE DES14398

IDLER BRACKET MODIFIED
TO CLEAR STARFISH F.S
CLUTCH DES 14387.

DRIVE TAKEN THROUGH
IDLER GEAR FROM
SPECIAL PINION ON PORT
SIDE F.S LAYSHAFT.

BASIC SCHEME DES 13125 AS INSTRUCTED FOR MERLIN Mᴋ 85

AUXILIARY GEARBOX DRIVE   MERLIN Mᴋ. 100.

FEB – 1944.
DRG REF Lᴅᴡ/Dᴡ/HP457

REF: M100/4071/

70

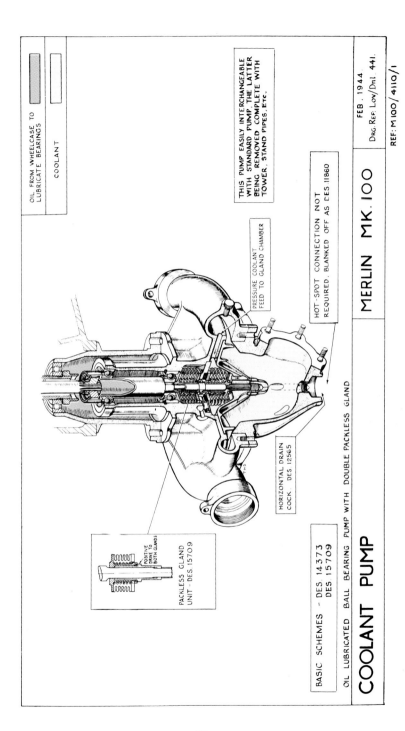

OIL FROM WHEELCASE TO
LUBRICATE BEARINGS

COOLANT

THIS PUMP EASILY INTERCHANGEABLE
WITH STANDARD PUMP, THE LATTER
BEING REMOVED COMPLETE WITH
TOWER, STAND PIPES, ETC.

PRESSURE COOLANT
FEED TO GLAND CHAMBER

HOT-SPOT CONNECTION NOT
REQUIRED. BLANKED OFF AS DES 11860

HORIZONTAL DRAIN
COCK. DES 12565

POSITIVE
DRIVE TO
BOTH GLANDS

PACKLESS GLAND
UNIT - DES 15709

BASIC SCHEMES - DES. 14373
DES. 15709

OIL LUBRICATED BALL BEARING PUMP WITH DOUBLE PACKLESS GLAND

COOLANT PUMP

MERLIN MK. 100

FEB. 1944.
DRG. REF. LOV/DNL. 441.

REF: M.100/4110/1

71

SLEEVE DOGS REDUCED IN DIA. BUT LENGTHENED. DES 16122

BACK PLATE DOGS REDUCED IN LENGTH TO ENABLE SEAL RING TO BE LAPPED DES 16112E

PUMP HOUSING UPPER FLANGE STRENGTHENED — LOWER FLANGE REDUCED. CASING MODIFIED TO FACILITATE DIE CASTING.
DES 16196

CIRCULAR BORE & SIX STUDS HOLDING CARBON RING HOUSING. DES 16369 ISSUE 2

THICKER CARBON RING & RECESS DEEPENED — ROTOR CONE ANGLE INCREASED FROM 33° TO 36°.
DES 15709 ISSUE 2

INTERNAL DIA OF SEAL RING INCREASED TO ALLOW ENTRY OF EXTENDED DOGS ON SLEEVE
DES 16122.

DRAWING SHEWING PUMP TO DES 15709

BASIC SCHEME DES 15709

COOLANT PUMP OIL LUBRICATED BALL BEARING ALL MERLINS

MAY ~ 1944
Lov/Dnl/HP 585.

MERLIN/4110/3

PUMP HOUSING UPPER FLANGE STRENGTHENED
LOWER FLANGE REDUCED - CASING MODIFED
TO FACILITATE DIE CASTING   DES. 16196

CIRCULAR BORE AND SIX STUDS
RETAINING CARBON RING HOUSING
DES. 16369 (ISS 2)

BACK-PLATE DOGS REDUCED IN LENGTH.
LONGER SLEEVE DOGS OF REDUCED DIAM.
INTERNAL DIAM OF SEAL RING INCREASED
DES 16122

CARBON RING HOUSING PARKERISED
INSTEAD OF NICKEL PLATED - TO
FACILITATE PRODUCTION

HOT-SPOT CONNECTION
BLANKED OFF ON MERLIN
100 SERIES

FOR DETAIL PART Nos AND FURTHER SCHEMES
SEE DRG REF: MERLIN/4110/6

NOV - 1944
DRG. REF. Lov/Dnl. 705

MERLIN/4110/4

BASIC SCHEME DES. 15709.

COOLANT PUMP   MERLIN 100 SERIES

COOLANT PUMP

MERLIN MK.100 SERIES

BASIC SCHEME DES. 15709

SPECIAL TOWER D.23743 MUST BE FITTED WITH THIS PUMP

HARDENED & STRENGTHENED SHAFT. & MODIFIED SPLINES, DES. 17011.

C 5113 SEEGER CIRCLIP

D 19892

GN 8975 (59 HOFFMAN BEARING)

D 26489 (155 HOFFMAN BEARING)

C 5128 SEEGER CIRCLIP

D 29017

K 4310
K 4506 } 4 off
K 8806
K 8 353

D 9462 } 2 off
D 2346

K 4304
K 4505
K 4446
D 2768/1 } 2 off
G 75908
K 3610

K 8 7253

D 23754 JOINTING

CR

D 28460 OUTLET CASING

DIE CAST CASING DES. 16196

D 2478

K 8 252 (6 off)

E 42876 (8 off)

D 23745 INLET CASING

E 42876 (6 off)

PROVISION OF DISTANCE PIECE DES. 17011.

D 13938

D 28437

D 28446 (PACKLESS GLAND ASSEMBLY)

K 8 8804
K 4506 } 6 off
K 4006

D 28198

D 28197

D 27263

THICKER CARBON RING DES. 15709 (ISSUE 2)

E 31212 (PLUG)

K 8 6110

E 24190

K 4404
K 4506 } 8 off
K 4006

D 29019

K 4156
K 8 7/104 } 6 off

D 28446 (PACKLESS GLAND ASSEMBLY)

D 28438/43

MODIFIED SLEEVE DES. 16122 DES. 17011

D 20372

E 46828
A 52279

COPPER WIRE 18 SWG x 6.000 LONG

D 19652/1

D 29018

D 15050 ASSEMBLY

E 16858

COPPER WIRE 20 SWG x 16.000 LONG

K 8 252

E 13709

D 7820

E 39704

K 4404

K 4506

K 4307

E 14748

ROTOR CONE ANGLE INCREASED FROM 33° TO 36° DES. 15709 (ISSUE 2)

GN 4680

K 8 7107

74

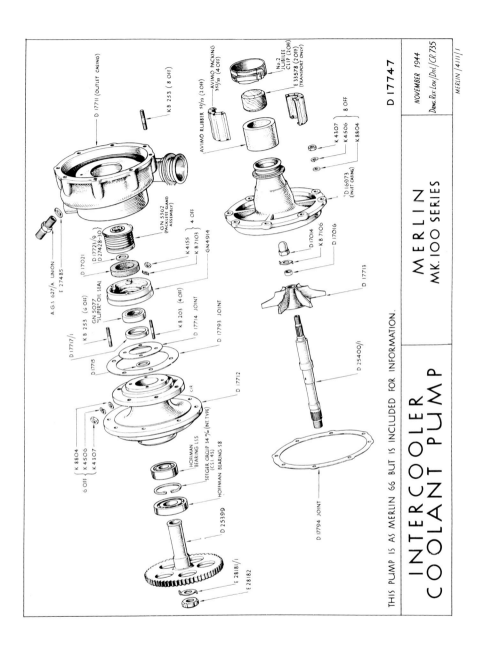

D 17747

NOVEMBER 1944

Dwg.Ref·Lov/Dnl/CR735

MERLIN/4111/1

MERLIN MK.100 SERIES

INTERCOOLER COOLANT PUMP

THIS PUMP IS AS MERLIN 66 BUT IS INCLUDED FOR INFORMATION.

D 17711 (OUTLET CASING)

K B 253 (8 OFF)

No.2 JUBILEE CLIP (2 OFF)

E 33578 (2 OFF) (TRANSPORT ONLY)

AVIMO PACKING 357/31 (4 OFF)

AVIMO RUBBER 57/31 (2 OFF)

K 4307 ⎱ 8 OFF
K 4506 ⎰
K 8804

D 16973 (INLET CASING)

K 4307
K 4506 ⎱ 6 OFF
K 8804 ⎰

GN 5362 (PACKLESS GLAND ASSEMBLY)

D 17723/9
D 27428-10 ⎱ 4 OFF

K 4155
K B 7101 ⎱ 4 OFF

GN 4914

A.G.1 627/A UNION

E 27485

D 17021

GN 5077 "SUPER" OIL SEAL

K B 253 (6 OFF)

K B 203 (4 OFF)

D 17714 JOINT

D 17793 JOINT

D 17717/1

D 17715

D 17712

CIR

K 8804
K 4506 ⎱ 6 OFF
K 4307 ⎰

HOFFMAN BEARING LS5

SEEGER CIRCLIP 34 3/4 (INT.TYPE) (CS.1.43)

HOFFMAN BEARING 58

D 25399

E 2818/1

E 28182

D 17014
K B 7106

D 17016

D 17713

D 25400/1

D 17794 JOINT

75

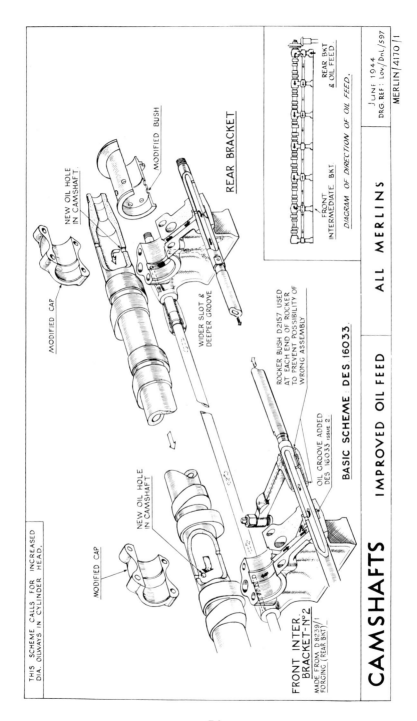

THIS SCHEME CALLS FOR INCREASED DIA. OILWAYS IN CYLINDER HEAD.

MODIFIED CAP

NEW OIL HOLE IN CAMSHAFT

MODIFIED BUSH

NEW OIL HOLE IN CAMSHAFT

MODIFIED CAP

WIDER SLOT & DEEPER GROOVE

REAR BRACKET

FRONT INTER. BRACKET-N°2
MADE FROM D.8239/1 FORGING (REAR BKT)

ROCKER BUSH D.2157 USED AT EACH END OF ROCKER TO PREVENT POSSIBILITY OF WRONG ASSEMBLY

OIL GROOVE ADDED DES 16033 issue 2

BASIC SCHEME: DES 16033

FRONT INTERMEDIATE BKT

REAR BKT & OIL FEED

DIAGRAM OF DIRECTION OF OIL FEED.

JUNE 1944
DRG REF: Lov/DnL/597

MERLIN/4170/1

CAMSHAFTS   IMPROVED OIL FEED   ALL MERLINS

76

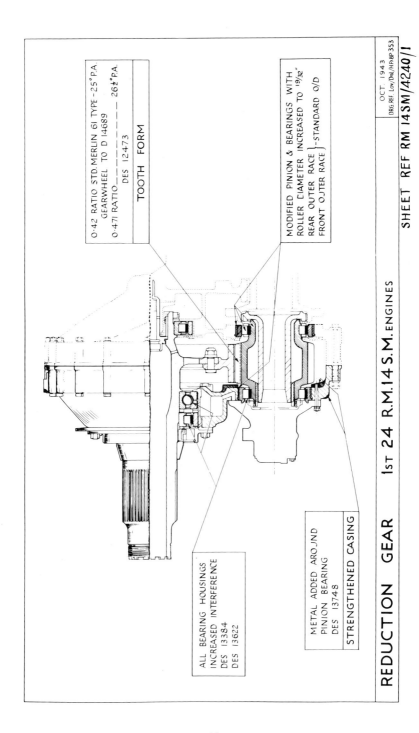

0·42 RATIO STD. MERLIN 61 TYPE - 25° P.A.
GEARWHEEL TO D 14689
0·471 RATIO ———————— 26½° P.A.
DES 12473

TOOTH FORM

MODIFIED PINION & BEARINGS WITH
ROLLER DIAMETER INCREASED TO ¹⁹/₃₂"
REAR OUTER RACE }-STANDARD O/D
FRONT OUTER RACE }

ALL BEARING HOUSINGS
INCREASED INTERFERENCE
DES 13384
DES 13622

METAL ADDED AROUND
PINION BEARING
DES 13748

STRENGTHENED CASING

REDUCTION GEAR   1ST 24 R.M.14 S.M. ENGINES

SHEET REF RM 14SM/4240/1

OCT. 1943
DRG. REF. Low/Dn/HR&P 353

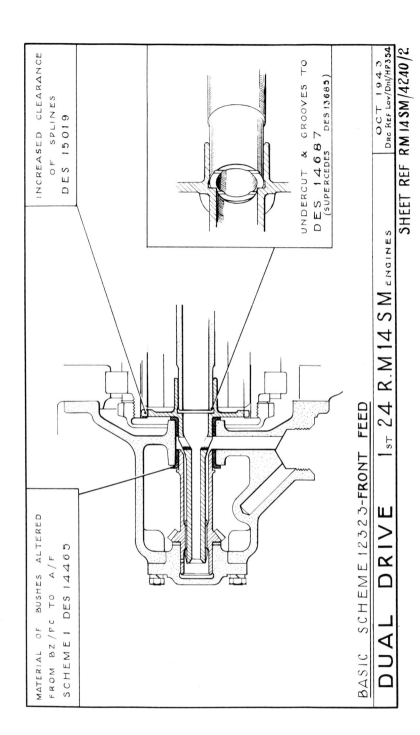

INCREASED CLEARANCE OF SPLINES DES 15019

UNDERCUT & GROOVES TO DES 14687 DES 13685
(SUPERCEDES DES 13685)

MATERIAL OF BUSHES ALTERED FROM BZ/PC TO A/F
SCHEME I DES 14465

BASIC SCHEME 12323-FRONT FEED 1ST 24 R.M 14 SM ENGINES

DUAL DRIVE

OCT 1943
DRG REF Lov/Dnl/HP 354

SHEET REF RM 14 SM/4240/2

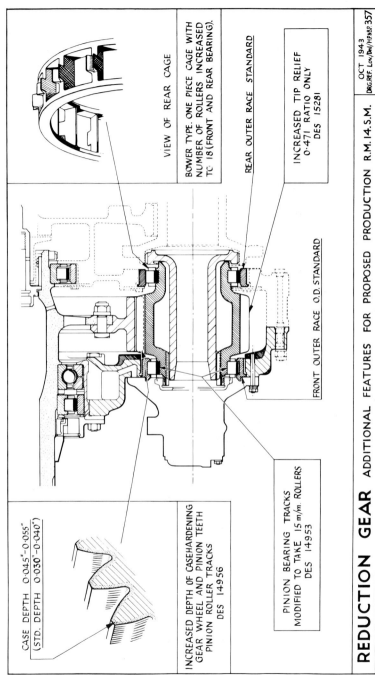

VIEW OF REAR CAGE

BOWER TYPE. ONE PIECE CAGE WITH NUMBER OF ROLLERS INCREASED TO 18 (FRONT AND REAR BEARING).

REAR OUTER RACE STANDARD

INCREASED TIP RELIEF 0·471 RATIO ONLY DES 15281

FRONT OUTER RACE O.D. STANDARD

PINION BEARING TRACKS MODIFIED TO TAKE 15 m/m. ROLLERS DES 14953

CASE DEPTH 0·045"–0·055" (STD. DEPTH 0·030"–0·040')

INCREASED DEPTH OF CASEHARDENING GEAR WHEEL AND PINION TEETH PINION ROLLER TRACKS DES 14956

REDUCTION GEAR ADDITIONAL FEATURES FOR PROPOSED PRODUCTION R.M. 14. S.M.

OCT 1943
DRG.REF. Lov./Dni/HP&&F 357

SHEET.REF: RM14SM/4240/3

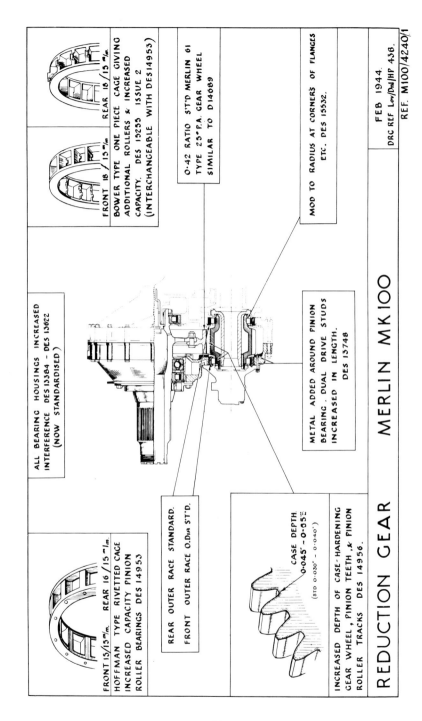

FRONT 18/15 m/m. REAR 18/15 m/m.
BOWER TYPE ONE PIECE CAGE GIVING ADDITIONAL ROLLERS & INCREASED CAPACITY. DES 15255 ISSUE 2 (INTERCHANGEABLE WITH DES14953)

0·42 RATIO ST'D MERLIN 61 TYPE 25° P.A. GEAR WHEEL SIMILAR TO D14689

MOD TO RADIUS AT CORNERS OF FLANGES ETC. DES 15532.

ALL BEARING HOUSINGS INCREASED INTERFERENCE DES 13384 - DES 13622 (NOW STANDARDISED)

METAL ADDED AROUND PINION BEARING. DUAL DRIVE STUDS INCREASED IN LENGTH. DES 13748

FRONT 15/15 m/m. REAR 16/15 m/m.
HOFFMAN TYPE RIVETTED CAGE INCREASED CAPACITY PINION ROLLER BEARINGS DES 14953

REAR OUTER RACE STANDARD. FRONT OUTER RACE O.Dia ST'D.

CASE DEPTH 0·045" - 0·05 ½"
(STD 0·030" - 0·040")

INCREASED DEPTH OF CASE-HARDENING GEAR WHEEL, PINION TEETH & PINION ROLLER TRACKS DES 14956.

REDUCTION GEAR MERLIN MK 100

FEB 1944.
DRG REF Lov/Dwl/HP 436.

REF. M100/4240/1

80

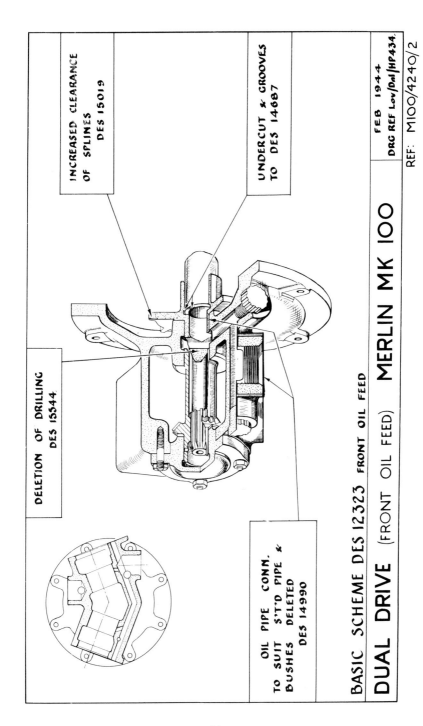

INCREASED CLEARANCE OF SPLINES DES 15019

UNDERCUT & GROOVES TO DES 14687

DELETION OF DRILLING DES 15544

OIL PIPE CONN. TO SUIT S'T'D PIPE & BUSHES DELETED DES 14990

BASIC SCHEME DES 12323 FRONT OIL FEED

DUAL DRIVE (FRONT OIL FEED) MERLIN MK 100

FEB 1944
DRG REF Lov/Dal/HP434.

REF: M100/4240/2

81

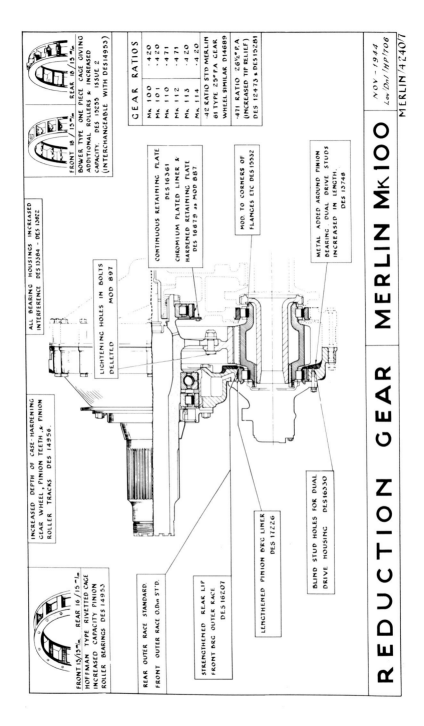

FRONT 15/13"m. REAR 16/15~"m.
HOFFMAN TYPE RIVETTED CAGE
INCREASED CAPACITY PINION
ROLLER BEARINGS DES 14953

REAR OUTER RACE STANDARD.
FRONT OUTER RACE O.Dꞩ STD.

STRENGTHENED REAR LIP
FRONT BRG OUTER RACE
DES 16207

LENGTHENED PINION BRG LINER
DES 17226

BLIND STUD HOLES FOR DUAL
DRIVE HOUSING   DES 16330

INCREASED DEPTH OF CASE-HARDENING
GEAR WHEEL, PINION TEETH & PINION
ROLLER TRACKS DES 14956.

ALL BEARING HOUSINGS INCREASED
INTERFERENCE DES 13384 - DES 13662

LIGHTENING HOLES IN BOLTS
DELETED   MOD 897

FRONT 18 /15"m   REAR 18 /13"m
BOWER TYPE ONE PIECE CAGE GIVING
ADDITIONAL ROLLERS & INCREASED
CAPACITY. DES 15255 ISSUE 2
(INTERCHANGEABLE WITH DES14953)

CONTINUOUS RETAINING PLATE
DES 16361

CHROMIUM PLATED LINER &
HARDENED RETAINING PLATE
DES 16679 & MOD 887

MOD. TO CORNERS OF
FLANGES ETC DES 15332

METAL ADDED AROUND PINION
BEARING. DUAL DRIVE STUDS
INCREASED IN LENGTH.
DES 13748

GEAR RATIOS	
Mk 100	.420
Mk 101	.420
Mk 110	.471
Mk 112	.471
Mk 113	.420
Mk 114	.420

.42 RATIO STD MERLIN
61 TYPE 25°PA GEAR
WHEEL SIMILAR D14689

.471 RATIO 26½°P.A.
(INCREASED TIP RELIEF)
DES 12473 & DES 15281

REDUCTION GEAR | MERLIN Mk100

NOV-1944
Lov/Dnl/HP706

MERLIN/42407

82

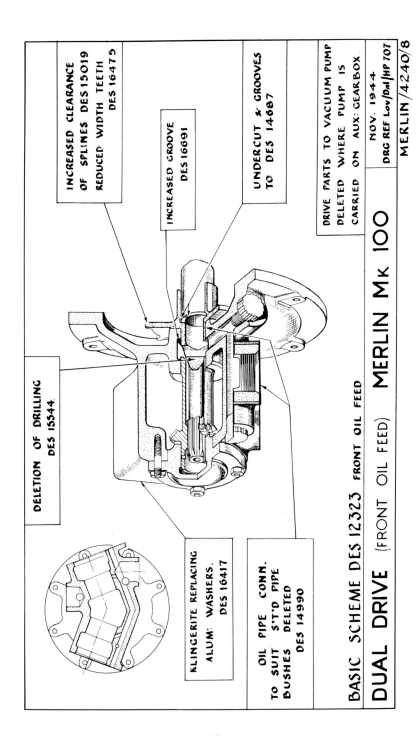

INCREASED CLEARANCE
OF SPLINES DES 15019
REDUCED WIDTH TEETH
DES 16475

INCREASED GROOVE
DES 16691

UNDERCUT & GROOVES
TO DES 14687

DRIVE PARTS TO VACUUM PUMP
DELETED WHERE PUMP IS
CARRIED ON AUX: GEARBOX.

DELETION OF DRILLING
DES 15544

KLINGERITE REPLACING
ALUM: WASHERS.
DES 16417

OIL PIPE CONN.
TO SUIT ST'D PIPE
BUSHES DELETED
DES 14990

BASIC SCHEME DES 12323 FRONT OIL FEED

DUAL DRIVE (FRONT OIL FEED) MERLIN MK 100

NOV. 1944.
DRG REF Lov/Dnl/HP 707

MERLIN/4240/8

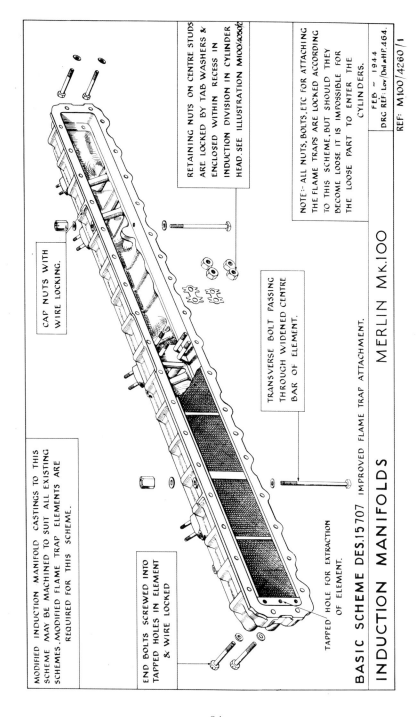

RETAINING NUTS ON CENTRE STUDS ARE LOCKED BY TAB WASHERS & ENCLOSED WITHIN RECESS IN INDUCTION DIVISION IN CYLINDER HEAD. SEE ILLUSTRATION M100/4050/2.

NOTE:- ALL NUTS, BOLTS, ETC FOR ATTACHING THE FLAME TRAPS ARE LOCKED ACCORDING TO THIS SCHEME, BUT SHOULD THEY BECOME LOOSE IT IS IMPOSSIBLE FOR THE LOOSE PART TO ENTER THE CYLINDERS.

FEB ~ 1944
DRG REF: Lov/Dml&HF.464.

REF: M100/4260/1

CAP NUTS WITH WIRE LOCKING.

TRANSVERSE BOLT PASSING THROUGH WIDENED CENTRE BAR OF ELEMENT.

MODIFIED INDUCTION MANIFOLD CASTINGS TO THIS SCHEME MAY BE MACHINED TO SUIT ALL EXISTING SCHEMES. MODIFIED FLAME TRAP ELEMENTS ARE REQUIRED FOR THIS SCHEME.

END BOLTS SCREWED INTO TAPPED HOLES IN ELEMENT & WIRE LOCKED

TAPPED HOLE FOR EXTRACTION OF ELEMENT.

BASIC SCHEME DES.15707 IMPROVED FLAME TRAP ATTACHMENT.

INDUCTION MANIFOLDS   MERLIN Mk.100

CAP NUTS WITH WIRE LOCKING

MANIFOLDS TO DES 15707 WITH CASTING MOD: TO DES 16180 MAY BE MACHINED TO SUIT THIS SCHEME.

NOTE: ALL NUTS, BOLTS, ETC FOR ATTACHING THE FLAME TRAPS ARE LOCKED ACCORDING TO THIS SCHEME. BUT SHOULD THEY BECOME LOOSE IT IS IMPOSSIBLE FOR THE LOOSE PART TO ENTER THE CYLINDERS.

RETAINING NUTS ON CENTRE STUDS ARE LOCKED BY TAB WASHER & ENCLOSED WITHIN INDUCTION PORT DIVISION (OR D-PIECE TO DES 14184) IN CYLINDER HEAD. WHERE INSERTS ARE FITTED THESE ARE REMOVED & A SPECIAL STUD FITTED.

END BOLT SCREWED INTO SPECIAL NUT RECESSED IN END OF ELEMENT & WIRE LOCKED

TRANSVERSE BOLT PASSING THROUGH CENTRE BAR OF ELEMENT & WIRE LOCKED.

BASIC SCHEME DES 16003 — IMPROVED FLAME TRAP ATTACHMENT SIMILAR TO DES 15707 BUT USING EXISTING MANIFOLDS & FLAME TRAP ELEMENTS.

INDUCTION MANIFOLDS — ALL MERLINS

MARCH — 1944
DRG REF: Low/Dnl/HP 505

REF: MERLIN/4260/2

85

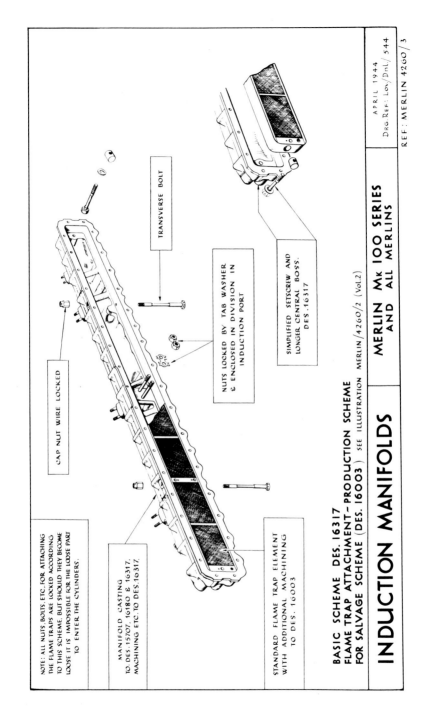

NOTE: ALL NUTS, BOLTS, ETC. FOR ATTACHING
THE FLAME TRAPS ARE LOCKED ACCORDING
TO THIS SCHEME, BUT SHOULD THEY BECOME
LOOSE IT IS IMPOSSIBLE FOR THE LOOSE PART
TO ENTER THE CYLINDERS.

CAP NUT WIRE LOCKED

TRANSVERSE BOLT

NUTS LOCKED BY TAB WASHER
& ENCLOSED IN DIVISION IN
INDUCTION PORT

SIMPLIFIED SETSCREW AND
LONGER CENTRAL BOSS.
DES. 16317

MANIFOLD CASTING
TO DES. 15707, 16180 & 16317,
MACHINING ETC. TO DES. 16317.

STANDARD FLAME TRAP ELEMENT
WITH ADDITIONAL MACHINING
TO DES. 16003

BASIC SCHEME DES. 16317
FLAME TRAP ATTACHMENT – PRODUCTION SCHEME
FOR SALVAGE SCHEME (DES. 16003) SEE ILLUSTRATION MERLIN/4260/2 (Vol.2)

MERLIN/4260/2 (Vol.2)

INDUCTION MANIFOLDS

MERLIN Mk 100 SERIES
AND ALL MERLINS

APRIL 1944
DRG. REF: Lov/Dnl/544

REF: MERLIN 4260/3

NOTE: ALL NUTS, BOLTS, ETC. FOR ATTACHING THE FLAME TRAPS ARE LOCKED ACCORDING TO THIS SCHEME, BUT SHOULD THEY BECOME LOOSE IT IS IMPOSSIBLE FOR THE LOOSE PART TO ENTER THE CYLINDERS.

CAP NUT WIRE LOCKED

4 BOSSES AND 4 END BOLT BOSSES DELETED. NEW POSITION OF ANCHOR LUGS DES.17359

END BOLT SCREWED INTO SPECIAL NUT RECESSED IN END OF ELEMENT AND WIRE LOCKED.

SIMPLIFIED SETSCREW & LONGER CENTRAL BOSS DES.16317

NUTS LOCKED BY TAB WASHER & ENCLOSED IN DIVISION IN INDUCTION PORT.

MANIFOLD CASTING TO DES. 15707, 16180 & 16317 MACHINING ETC. TO DES.16317

STANDARD FLAME TRAP ELEMENT WITH ADDITIONAL MACHINING TO DES. 16003

BASIC SCHEME DES.16317 FLAME TRAP ATTACHMENT — PRODUCTION SCHEME

INDUCTION MANIFOLDS | MERLIN Mk 100 SERIES

NOV.1944
DRG.REF:Lov./DnL/708

MERLIN/4260/4

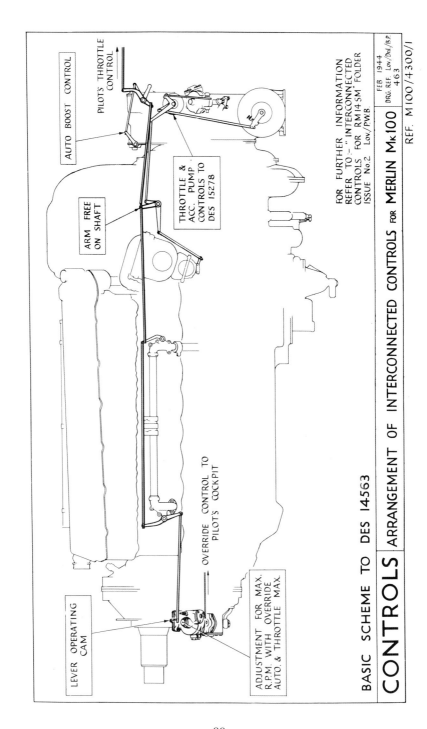

PILOT'S THROTTLE CONTROL

AUTO BOOST CONTROL

ARM FREE ON SHAFT

THROTTLE & ACC. PUMP CONTROLS TO DES 15278

LEVER OPERATING CAM

OVERRIDE CONTROL TO PILOT'S COCKPIT

ADJUSTMENT FOR MAX. R.P.M. WITH OVERRIDE AUTO. & THROTTLE MAX.

FOR FURTHER INFORMATION
REFER TO:- " INTERCONNECTED
CONTROLS FOR RM14 SM" FOLDER
ISSUE No.2   Lov/PWB

FEB 1944
DRG. REF. Lov./Dnl./B.P.
463

BASIC SCHEME TO DES 14563

CONTROLS ARRANGEMENT OF INTERCONNECTED CONTROLS FOR MERLIN MK100

REF. M100/4300/1

88

THROTTLE & ACC. PUMP CONTROL
DES. 15278

THROTTLE TORQUE SPRING BALANCE
DES. 16888

MODIFIED MAGNETO
CONTROL LEVER.
CASING FLANGES
DES. 15833

VIEW ON STARBOARD SIDE

NEW INDEX PLATE
DES. 17283

BASIC SCHEME TO SUIT S.U. INJECTION SYSTEM DES. 15278

CONTROLS   MERLIN Mk. 100

NOV. 1944
DRG. REF: Lov/Dnl/CIR 709

MERLIN/4300/2

FUEL PUMP DRIVE ·9167 × ENGINE SPEED
FOR S. U. PUMP. DES. 14179.

DRIVING GEAR

MAIN DRIVE SHAFT

SPUR GEAR FEED PUMP

SERVO PISTON

SWASH PLATE

"Z" SHAFT

PLUNGER

INLET FROM FEED PUMP

ECCENTRIC DISTRIBUTOR VALVE

SLOW RUNNING CUT OFF

FOR FURTHER INFORMATION REFER TO:-
"THE SINGLE-POINT S.U. INJECTION PUMP"
FOLDER, ISSUE No 3, JAN 1944.

FEB. 1944.
DRG. REF. Lav/Dhl/BP 451

REF: M100/4361/1

MERLIN MK.100.

FUEL OUTLET

INLET

BOUGHT OUT COMPLETE

FUEL INJECTOR   S.U. FUEL INJECTION PUMP

90

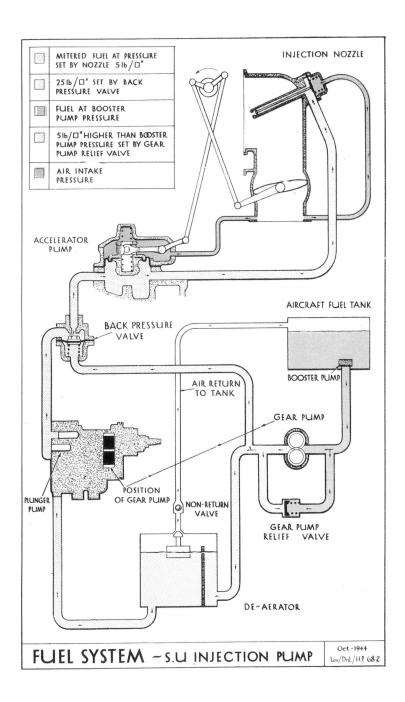

METERED FUEL AT PRESSURE
SET BY NOZZLE 5 lb /$\square$"

25 lb /$\square$" SET BY BACK
PRESSURE VALVE

FUEL AT BOOSTER
PUMP PRESSURE

5 lb /$\square$" HIGHER THAN BOOSTER
PUMP PRESSURE SET BY GEAR
PUMP RELIEF VALVE

AIR INTAKE
PRESSURE

INJECTION NOZZLE

ACCELERATOR
PUMP

AIRCRAFT FUEL TANK

BACK PRESSURE
VALVE

AIR RETURN
TO TANK

BOOSTER PUMP

GEAR PUMP

PLUNGER
PUMP

POSITION
OF GEAR PUMP

NON-RETURN
VALVE

GEAR PUMP
RELIEF VALVE

DE-AERATOR

FUEL SYSTEM – S.U INJECTION PUMP

Oct -1944
Lov/Dnl /H.P. 68.2

91

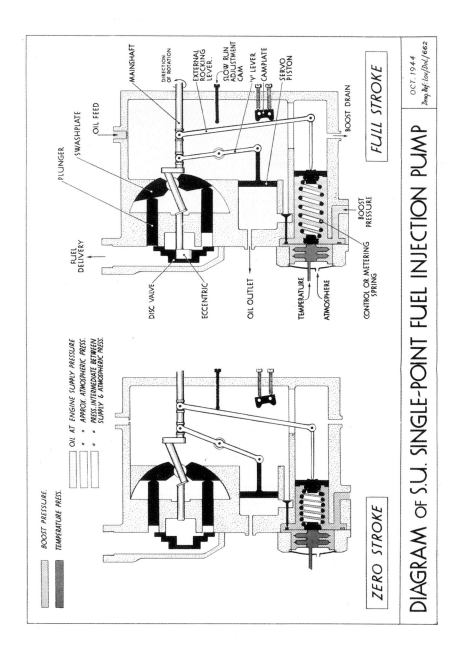

FULL STROKE

OCT. 1944

Drng.Ref-Lou/Dal/662

MAINSHAFT
DIRECTION OF ROTATION
EXTERNAL ROCKING LEVER.
SLOW RUN ADJUSTMENT CAM
'Y' LEVER
CAMPLATE
SERVO PISTON
OIL FEED
SWASHPLATE
PLUNGER
BOOST DRAIN
BOOST PRESSURE
FUEL DELIVERY
DISC VALVE
ECCENTRIC
OIL OUTLET
TEMPERATURE
ATMOSPHERE
CONTROL OR METERING SPRING

ZERO STROKE

BOOST PRESSURE.
TEMPERATURE PRESS.

OIL AT ENGINE SUPPLY PRESSURE
" = APPROX. ATMOSPHERIC PRESS.
" = PRESS. INTERMEDIATE BETWEEN SUPPLY & ATMOSPHERIC PRESS.

DIAGRAM OF S.U. SINGLE-POINT FUEL INJECTION PUMP

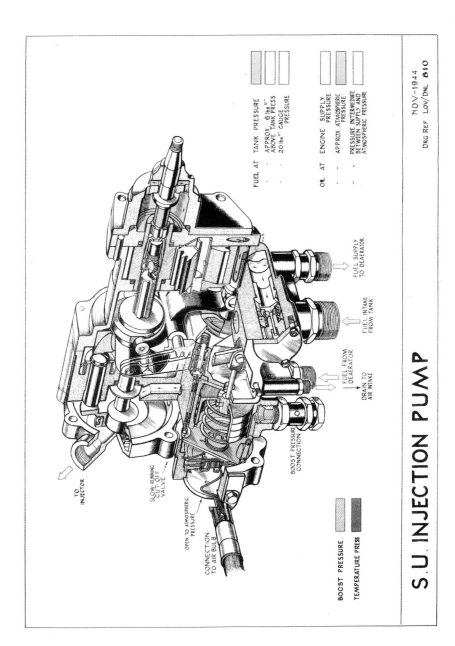

S.U. INJECTION PUMP

TO INJECTOR

SLOW-RUNNING CUT-OFF VALVE

OPEN TO ATMOSPHERIC PRESSURE

CONNECTION TO AIR BULB

BOOST PRESSURE CONNECTION

DRAIN TO AIR INTAKE

FUEL FROM DEAERATOR

FUEL INTAKE FROM TANK

FUEL SUPPLY TO DEAERATOR

FUEL AT TANK PRESSURE

„ „ APPROX. 6 lbs □" ABOVE TANK PRESS

„ „ 20 lbs □" GAUGE PRESSURE

OIL AT ENGINE SUPPLY PRESSURE

„ „ APPROX. ATMOSPHERIC PRESSURE

„ „ PRESSURE INTERMEDIATE BETWEEN SUPPLY AND ATMOSPHERIC PRESSURE.

BOOST PRESSURE.

TEMPERATURE PRESS

NOV-1944.
DRG REF LOV/DNL. 810

93

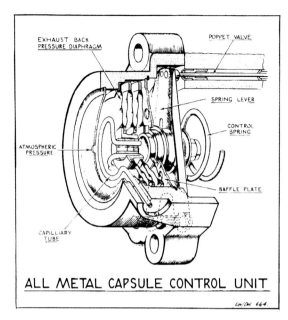

ALL METAL CAPSULE CONTROL UNIT

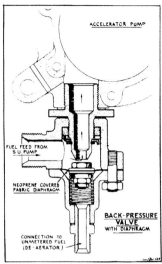

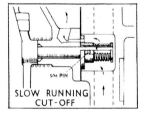

SLOW RUNNING CUT-OFF

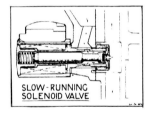

SLOW-RUNNING SOLENOID VALVE

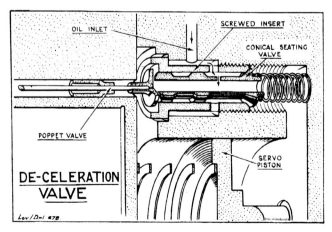

DE-CELERATION VALVE

Components of the S.U. Fuel Injection Pump

94

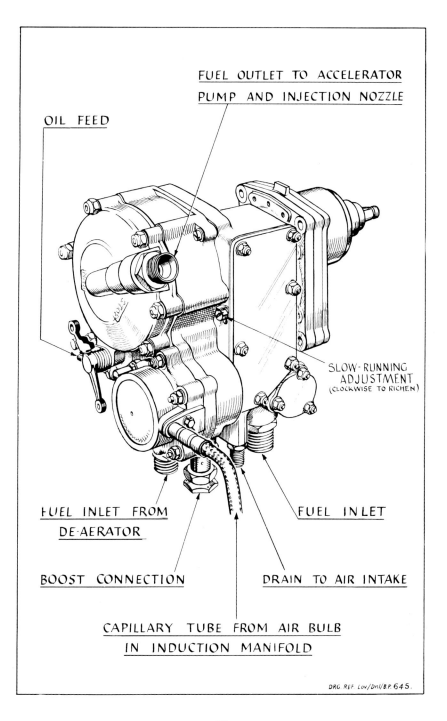

FUEL OUTLET TO ACCELERATOR
PUMP AND INJECTION NOZZLE

OIL FEED

SLOW-RUNNING
ADJUSTMENT
(CLOCKWISE TO RICHEN)

FUEL INLET FROM
DE-AERATOR

FUEL INLET

BOOST CONNECTION

DRAIN TO AIR INTAKE

CAPILLARY TUBE FROM AIR BULB
IN INDUCTION MANIFOLD

DRG. REF. Lov/Dni/B.P. 645.

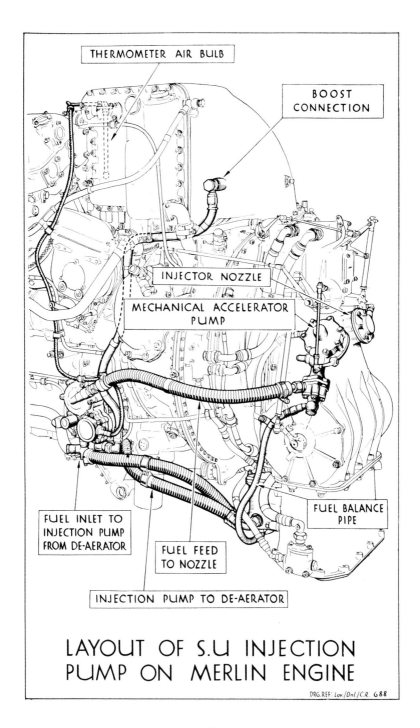

THERMOMETER AIR BULB

BOOST CONNECTION

INJECTOR NOZZLE

MECHANICAL ACCELERATOR PUMP

FUEL BALANCE PIPE

FUEL INLET TO INJECTION PUMP FROM DE-AERATOR

FUEL FEED TO NOZZLE

INJECTION PUMP TO DE-AERATOR

LAYOUT OF S.U INJECTION PUMP ON MERLIN ENGINE

DRG.REF: Lov./Dnl/C.R. 688

NON RETURN BALL VALVE. DES 14239

FUEL OUTLET

PIPE RUNS FOR DE-AERATOR / S.U. PUMP AND VOLUTE DRAIN. DES 16596.

VAPOUR RETURN TO TANK.

MODIFIED ALTERNATIVE POSITION OF FUEL INLET DES 16663

FUEL CONNECTIONS DES 16682

MOUNTING OF DE-AERATOR ON AIR INTAKE ADAPTOR DES 16485 (ISSUE 2)

OUTLET TO B.P. VALVE.

FUEL BALANCE PIPE. DES 7640

INJECTION PUMP TO DE-AERATOR

FUEL RETURN FROM DE-AERATOR TO INJECTION PUMP

BASIC SCHEME DES 14000 (ISSUE 2)

DE-AERATOR

MERLIN 100

DEC. 1944
DRG. REF: Lov/Dml/B.P. 769

MERLIN / 4364 / 1

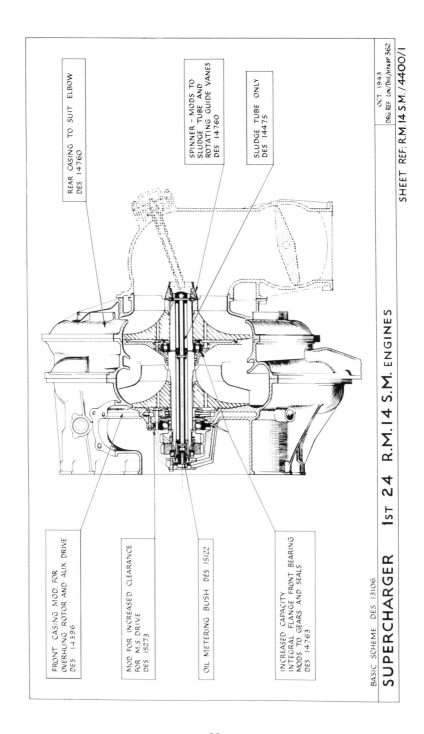

REAR CASING TO SUIT ELBOW
DES 14760

SPINNER – MODS TO
SLUDGE TUBE AND
ROTATING GUIDE VANES
DES 14760

SLUDGE TUBE ONLY
DES 14475

FRONT CASING MOD FOR
OVERHUNG ROTOR AND AUX. DRIVE
DES 14396

MOD. FOR INCREASED CLEARANCE
FOR M.S. DRIVE
DES 15273

OIL METERING BUSH DES 15122

INCREASED CAPACITY
INTEGRAL FLANGE FRONT BEARING
MODS TO GEARS AND SEALS
DES 14763

BASIC SCHEME DES 13106

SUPERCHARGER 1st 24 R.M.14 S.M. ENGINES

OCT. 1943
DRG. REF. Low/Dril/HF&BP 362

SHEET REF. R.M.14 S.M./4400/1

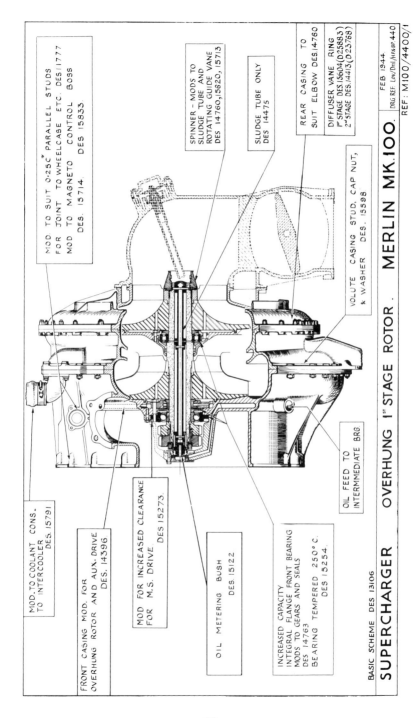

MOD. TO COOLANT CONS.
TO INTERCOOLER
DES. 15791

MOD. TO SUIT 0·25" PARALLEL STUDS
FOR JOINT TO WHEELCASE ETC. DES. 11777
MOD TO MAGNETO CONTROL BOSS
DES. 15714.   DES. 15853

FRONT CASING MOD. FOR
OVERHUNG ROTOR AND AUX. DRIVE
DES. 14396

SPINNER – MODS TO
SLUDGE TUBE AND
ROTATING GUIDE VANE
DES. 14760, 15620, 15713

SLUDGE TUBE ONLY
DES. 14475

REAR CASING TO
SUIT ELBOW DES. 14760

DIFFUSER VANE RING
1ˢᵗ STAGE DES. 15604 (D25883)
2ⁿᵈ STAGE DES. 14413, (D23768)

MOD FOR INCREASED CLEARANCE
FOR M.S. DRIVE
DES 15273.

OIL METERING BUSH
DES. 15122

VOLUTE CASING STUD, CAP NUT,
& WASHER DES. 15598

INCREASED CAPACITY
INTEGRAL FLANGE FRONT BEARING
MODS TO GEARS AND SEALS
DES 14763
BEARING TEMPERED 250° C.
DES 15254.

OIL FEED TO
INTERMMEDIATE BRG

BASIC SCHEME DES 13106

SUPERCHARGER   OVERHUNG 1ˢᵗ STAGE ROTOR.   MERLIN MK.100.

FEB 1944.
DRG REF. Low/Dril/Hrs BP 440

REF: M100/4400/1

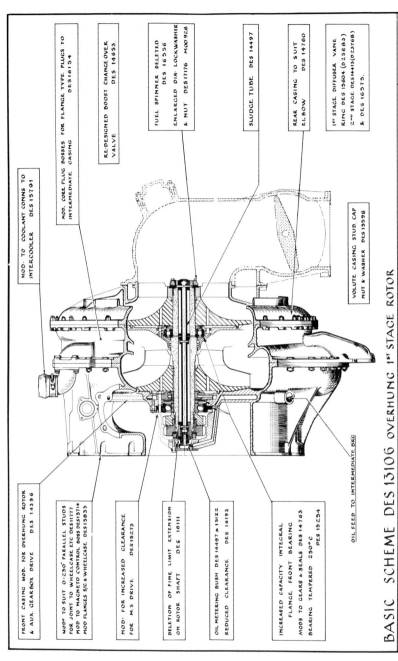

MOD. TO COOLANT CONNS TO INTERCOOLER DES 15791

MOD. CORE PLUG BOSSES FOR FLANGE TYPE PLUGS TO INTERMEDIATE CASING DES 16134

RE-DESIGNED BOOST CHANGE OVER VALVE DES 14653

FUEL SPINNER DELETED DES 16556
ENLARGED DIA. LOCKWASHER & NUT DES17176 MOD 926

SLUDGE TUBE DES 14497

REAR CASING TO SUIT ELBOW DES 14760

1ST STAGE DIFFUSER VANE RING DES 15604 (025883)
2ND STAGE DES14413(023768) & DES 16375.

VOLUTE CASING STUD CAP NUT & WASHER DES 15598

OIL FEED TO INTERMEDIATE DRG

FRONT CASING MOD. FOR OVERHUNG ROTOR & AUX. GEARBOX DRIVE DES 14396

MOD? TO SUIT 0.250" PARALLEL STUDS FOR JOINT TO WHEELCASE ETC DES11777
MOD TO MAGNETO CONTROL BOSS DES13714
MOD FLANGES S/C & WHEELCASE DES15833

MOD? FOR INCREASED CLEARANCE FOR M.S DRIVE. DES15273

DELETION OF FINE LIMIT EXTENSION ON ROTOR SHAFT DES 16111

OIL METERING BUSH DES 14497 & 15122
REDUCED CLEARANCE DES 16193

INCREASED CAPACITY INTEGRAL FLANGE FRONT BEARING
MODS TO GEARS & SEALS DES14763
BEARING TEMPERED 250°C DES 15254

BASIC SCHEME DES 13106 OVERHUNG 1ST STAGE ROTOR

SUPERCHARGER | MERLIN MK 100

NOV - 1944
Lov/Dal/MP/712

MERLIN/4400/5

BOSSES FOR CHANGE-OVER VALVE
FOR R.R. INJECTOR PUMP ONLY
DES 15119

INJECTOR BODY
MECH. ACC. PUMP
DES 14970

½" PETROL CONNECTION
FOR S.U. PUMP
DES 15096

NOZZLE PIPE
DES 15169

SHORT ELBOW
SINGLE PLATE THROTTLE
DES 14760

S/C. INTAKE. 1ST 24 R.M.14 S.M. ENGINES

OCT. 1943
DRG. REF. Low/Dni/Mpl&P 363

SHEET REF: R M 14 S.M. 4406/1

BOSSES FOR CHANGE OVER VALVE
FOR R.R INJECTOR PUMP ONLY
DES 15119

INJECTOR BODY
MECH. ACC. PUMP
DES 14,970

¼" PETROL CONNECTION
FOR S.U. PUMP
DES 15096

INTAKE GAUZE OF STRIP
CONSTRUCTION - DES 15574

NOZZLE PIPE
DES 15169

SHEET METAL VANE
ADDED - DES 15593

SHORT ELBOW
SINGLE PLATE THROTTLE
DES 14760

S/C. INTAKE. 1ST 24 R.M.14 S.M. ENGINES

OCT. 1943
DRG. REF. Lov. Dhl/MF **370**

SHEET REF: R.M.I.4 S.M. 44 OG/2

102

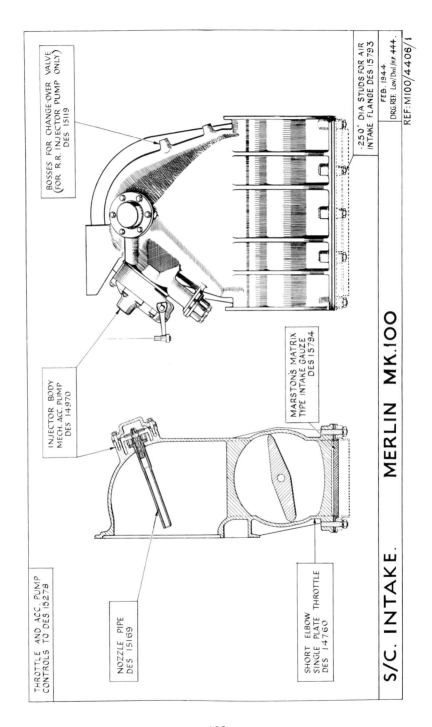

THROTTLE AND ACC. PUMP
CONTROLS TO DES 15278

BOSSES FOR CHANGE-OVER VALVE
(FOR R.R. INJECTOR PUMP ONLY)
DES 15119

.250" DIA STUDS FOR AIR
INTAKE FLANGE DES 15793

FEB.1944
DRG.REF. Low/DHI/HP. 444.

REF: M100/4406/1

INJECTOR BODY
MECH. ACC. PUMP
DES 14970

MARSTON'S MATRIX
TYPE INTAKE GAUZE
DES 15794

NOZZLE PIPE
DES 15169

SHORT ELBOW
SINGLE PLATE THROTTLE
DES 14760

S/C. INTAKE.   MERLIN MK.100

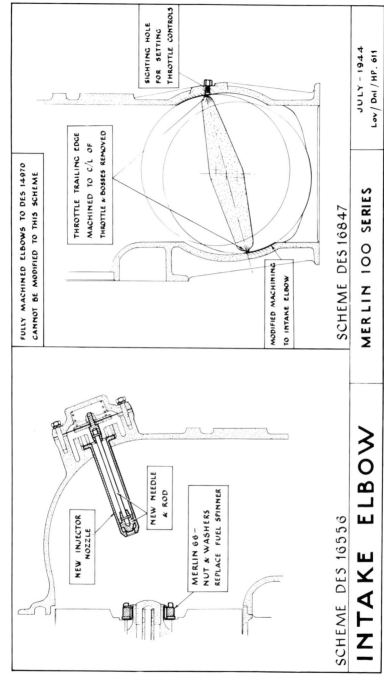

FULLY MACHINED ELBOWS TO DES 14970 CANNOT BE MODIFIED TO THIS SCHEME

SIGHTING HOLE FOR SETTING THROTTLE CONTROLS

THROTTLE TRAILING EDGE MACHINED TO C/L OF THROTTLE & BOSSES REMOVED

MODIFIED MACHINING TO INTAKE ELBOW

SCHEME DES 16847

MERLIN 100 SERIES

NEW INJECTOR NOZZLE

NEW NEEDLE & ROD

MERLIN 66 – NUT & WASHERS REPLACE FUEL SPINNER

SCHEME DES 16556

INTAKE ELBOW

JULY - 1944
Lov/Dnl/HP. 611

MERLIN/4406/5

BOSSES FOR
CHANGE-OVER
VALVE
(FOR R.R. INJECTOR
PUMP ONLY)
DES. 15119

AIR INTAKE
ADAPTOR AND
VOLUTE DRAIN
DES. 16485
(ISSUE 2)

CIR

MOD INJECTION
NOZZLE – DES. 17612

DOUBLE DIAPHRAGM
ACCELERATOR PUMP
DES. 16647

BACK PRESSURE
VALVE
DES. 16656

SIGHTING HOLE
FOR SETTING
THROTTLE CONTROLS
DES. 16847
TAB WASH. LOCKING
DES. 17010

STRENGTHENED DOUBLE DIAPHRAGM TO DES. 17227

INJECTION NOZZLE
NEEDLE & ROD
DES. 16556

SHORT ELBOW
SINGLE PLATE
THROTTLE DES. 14760
MACHINING MOD.
TO THROTTLE &
ELBOW DES. 16847

THROTTLE & ACC. PUMP CONTROLS DES. 15278

BASIC SCHEME DES. 14760   S/C INTAKE   MERLIN Mk. 100

NOV. 1944
DRG. REF. Low/DnL/CIR 713

MERLIN/4406/6

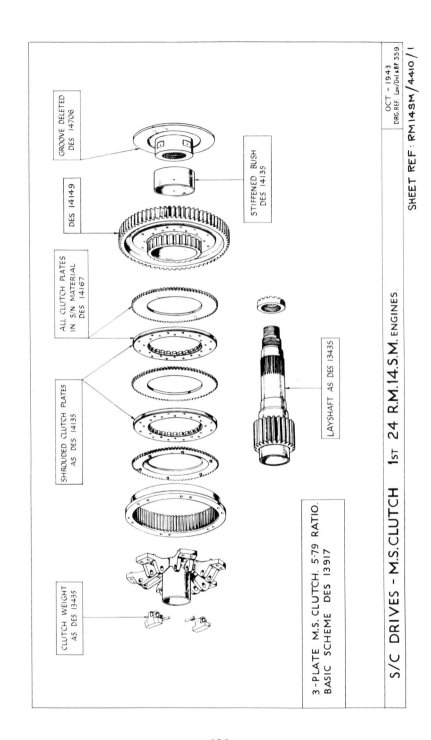

GROOVE DELETED
DES 14706

DES 14149

STIFFENED BUSH
DES 14135

ALL CLUTCH PLATES
IN S/N MATERIAL
DES 14167

SHROUDED CLUTCH PLATES
AS DES 14135

LAYSHAFT AS DES 13435

CLUTCH WEIGHT
AS DES 13435

3-PLATE M.S. CLUTCH. 5·79 RATIO.
BASIC SCHEME DES 13917

S/C DRIVES - M.S. CLUTCH 1ST 24 R.M.14.S.M. ENGINES.

OCT - 1943
DRG. REF. Low/Dwl & BF 359.

SHEET REF: RM14SM/4410/1

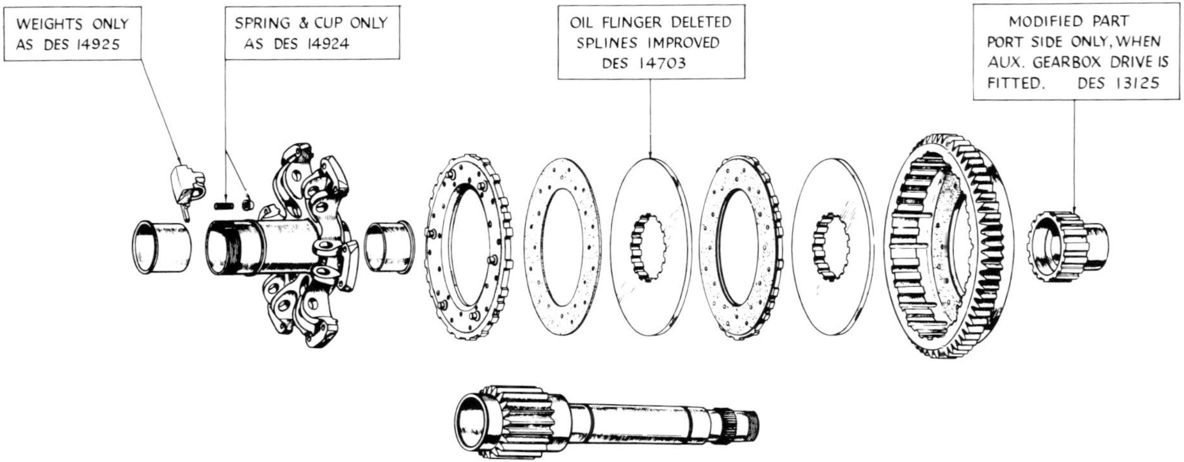

WEIGHTS ONLY
AS DES 14925
SPRING & CUP ONLY
AS DES 14924
OIL FLINGER DELETED
SPLINES IMPROVED
DES 14703
MODIFIED PART
PORT SIDE ONLY, WHEN
AUX. GEARBOX DRIVE IS
FITTED. DES 13125

STAR CARRIER PLATE
F.S. CLUTCH. 7·06 RATIO
BASIC SCHEME DES 14151

S/C DRIVE - F.S. CLUTCH 1ST 24 R.M.14.S.M. ENGINES

OCT. 1943
DRG. REF. Lov/Dnl. 358

SHEET REF RM 14SM/4410/2

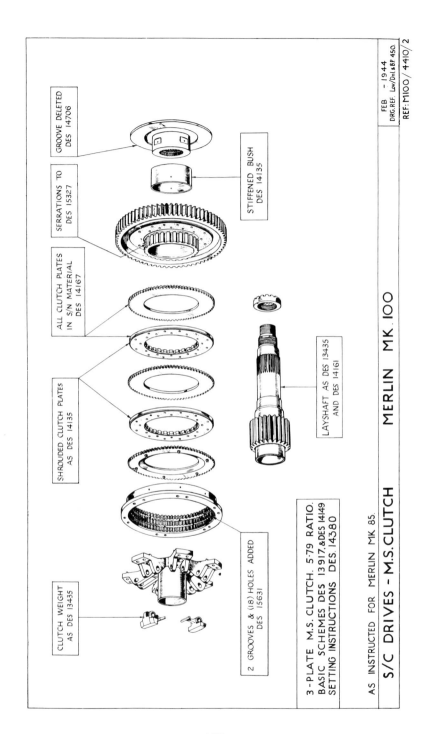

GROOVE DELETED
DES 14706

SERRATIONS TO
DES 15327

STIFFENED BUSH
DES 14135

ALL CLUTCH PLATES
IN S/N MATERIAL
DES 14167

SHROUDED CLUTCH PLATES
AS DES 14135

LAYSHAFT AS DES 13435
AND DES 14161

CLUTCH WEIGHT
AS DES 13435

2 GROOVES & (18) HOLES ADDED
DES 15631

3-PLATE M.S.CLUTCH. 5·79 RATIO.
BASIC SCHEMES DES 13917.& DES 14149
SETTING INSTRUCTIONS DES.14380

AS INSTRUCTED FOR MERLIN MK 85.

S/C DRIVES - M.S.CLUTCH   MERLIN MK.100

FEB - 1944
DRG.REF. Lav/Dml & Bp 450.

REF: M100 / 4410/ 2

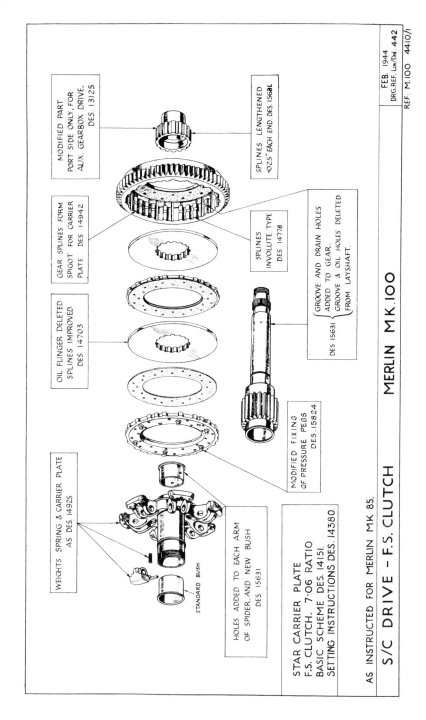

MODIFIED PART
PORT SIDE ONLY, FOR
AUX. GEARBOX DRIVE.
DES. 13125

SPLINES LENGTHENED
·025" EACH END. DES. 15631

GEAR SPLINES FORM
SPIGOT FOR CARRIER
PLATE. DES. 14942

SPLINES
INVOLUTE TYPE
DES. 14778

GROOVE AND DRAIN HOLES
ADDED TO GEAR.
GROOVE & OIL HOLES DELETED
FROM LAYSHAFT.
DES. 15631

OIL FLINGER DELETED
SPLINES IMPROVED
DES. 14703

MODIFIED FIXING
OF PRESSURE PEGS
DES. 15824

WEIGHTS SPRING & CARRIER PLATE
AS DES. 14925

STANDARD BUSH

HOLES ADDED TO EACH ARM
OF SPIDER, AND NEW BUSH
DES. 15631

STAR CARRIER PLATE
F.S. CLUTCH. 7·06 RATIO
BASIC SCHEME DES. 14151.
SETTING INSTRUCTIONS DES. 14380.

AS INSTRUCTED FOR MERLIN MK 85.

MERLIN MK. 100

S/C DRIVE – F.S. CLUTCH

FEB. 1944
DRG.REF. Lox/Dwl 442

REF: M.100 44I0/I

109

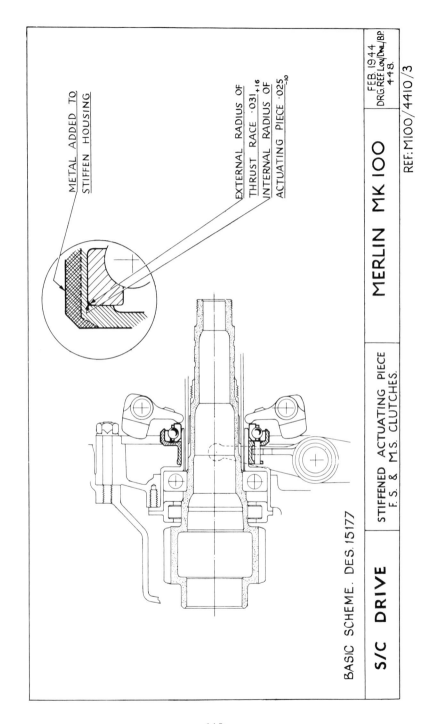

METAL ADDED TO STIFFEN HOUSING

EXTERNAL RADIUS OF THRUST RACE ·031 +·16
INTERNAL RADIUS OF ACTUATING PIECE ·025 -·10

BASIC SCHEME. DES. 15177

STIFFENED ACTUATING PIECE F. S. & M.S. CLUTCHES.

MERLIN MK 100

S/C DRIVE

FEB. 1944
DRG. REF. Lov/Dnl/BP. 448.

REF: M100/4410/3

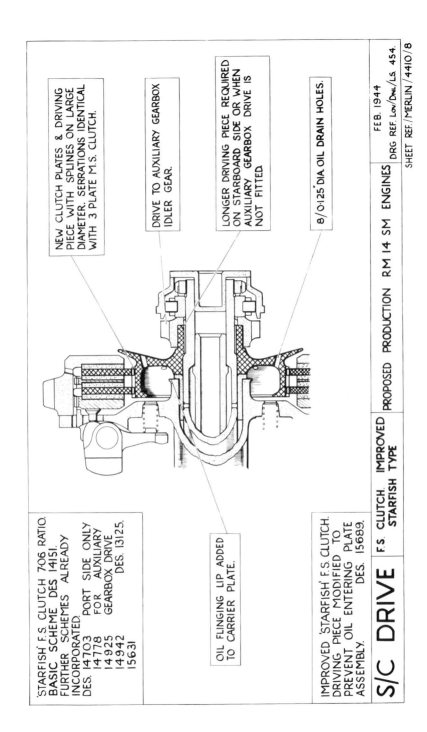

NEW CLUTCH PLATES & DRIVING PIECE WITH SPLINES ON LARGE DIAMETER. SERRATIONS IDENTICAL WITH 3 PLATE M.S. CLUTCH.

DRIVE TO AUXILIARY GEARBOX IDLER GEAR.

LONGER DRIVING PIECE REQUIRED ON STARBOARD SIDE OR WHEN AUXILIARY GEARBOX DRIVE IS NOT FITTED

8/0·125" DIA OIL DRAIN HOLES.

'STARFISH' F.S. CLUTCH 706 RATIO. BASIC SCHEME DES 14151. FURTHER SCHEMES ALREADY INCORPORATED.
DES. 14703   PORT SIDE ONLY
14778   FOR AUXILIARY
14925   GEARBOX DRIVE
14942   DES. 13125.
15631

OIL FLINGING LIP ADDED TO CARRIER PLATE.

IMPROVED 'STARFISH' F.S. CLUTCH. DRIVING PIECE MODIFIED TO PREVENT OIL ENTERING PLATE ASSEMBLY.   DES. 15689.

S/C DRIVE

F.S. CLUTCH. IMPROVED STARFISH TYPE	PROPOSED PRODUCTION RM 14 SM ENGINES

FEB. 1944
DRG REF. Lov/Dwl/LS. 454.

SHEET REF./MERLIN/4410/8

111

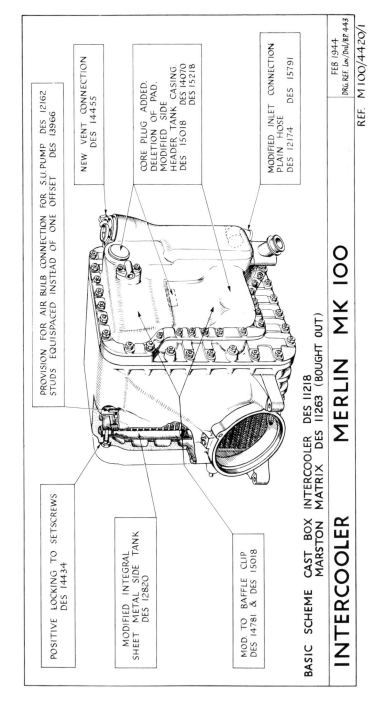

PROVISION FOR AIR BULB CONNECTION FOR S.U. PUMP DES 12162
STUDS EQUISPACED INSTEAD OF ONE OFFSET DES 13966

NEW VENT CONNECTION
DES 14455

CORE PLUG ADDED.
DELETION OF PAD.
MODIFIED SIDE
HEADER TANK CASING.
DES 15018   DES 14070
            DES 15218

MODIFIED INLET CONNECTION
PLAIN HOSE
DES 12174   DES 15791

POSITIVE LOCKING TO SETSCREWS
DES 14434

MODIFIED INTEGRAL
SHEET METAL SIDE TANK
DES 12820

MOD. TO BAFFLE CLIP
DES 14781 & DES 15018

BASIC SCHEME CAST BOX INTERCOOLER DES 11218
MARSTON MATRIX DES 11263 (BOUGHT OUT)

INTERCOOLER MERLIN MK 100

FEB 1944
DRG. REF. Lon/Dnl/BP. 443

REF. M 100/4420/1

NEW VENT CONNECTION
DES 14455

ALUM. JOINT WASHER
ADDED. DES. 16993.

CORE PLUG ADDED. DELETION
OF PAD. MODIFIED SIDE
HEADER TANK CASING.
DES 15018 14070, & 15214

DELETION OF VANES
IN INTR REAR ELBOW
DES.16109

MODIFIED INLET CONNECTION
PLAIN HOSE
DES 12174   DES 15791

PROVISION FOR AIR BULB CONNECTION FOR S.U.PUMP DES 12162
STUDS EQUISPACED INSTEAD OF ONE OFFSET   DES 13966

MOD. SANDWICH-PLATE. DES.15797

BOSS ADDED FOR VENT
PIPE CLIP. DES. 16223.

POSITIVE LOCKING TO SETSCREWS
DES 14434

MODIFIED INTEGRAL
SHEET METAL SIDE TANK
DES 12820

MOD. TO BAFFLE CLIP
DES 14781 & DES 15018

MOD. BOLTS TO BRIDGE
PIECE SUPPORT.
DES.16366 (MOD. 827)

BASIC SCHEME CAST BOX INTERCOOLER DES 11218
MARSTON MATRIX DES 11263 (BOUGHT OUT)

INTERCOOLER   MERLIN Mk.100

NOV.1944
Drg.Ref:Lov/Dnl/B.P. 714

MERLIN/4420/2

113

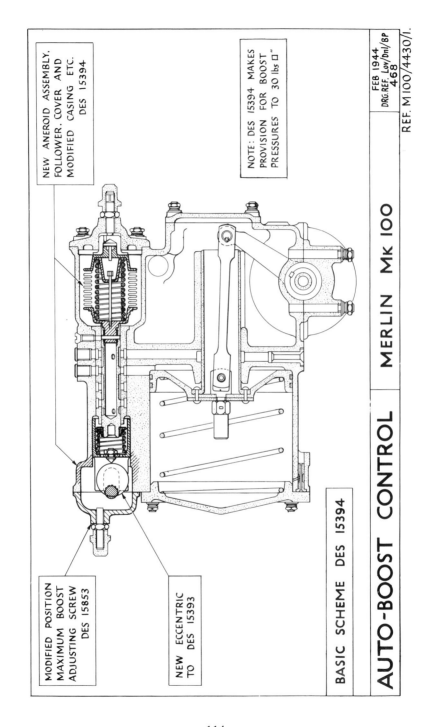

NEW ANEROID ASSEMBLY,
FOLLOWER, COVER AND
MODIFIED CASING ETC.
DES 15394

NOTE: DES 15394 MAKES
PROVISION FOR BOOST
PRESSURES TO 30 lbs □"

MODIFIED POSITION
MAXIMUM BOOST
ADJUSTING SCREW
DES 15853

NEW ECCENTRIC
TO DES 15393

BASIC SCHEME DES 15394

AUTO-BOOST CONTROL | MERLIN Mk 100

FEB 1944
DRG.REF. Lav/bnl/BP
468
REF. M100/4430/1.

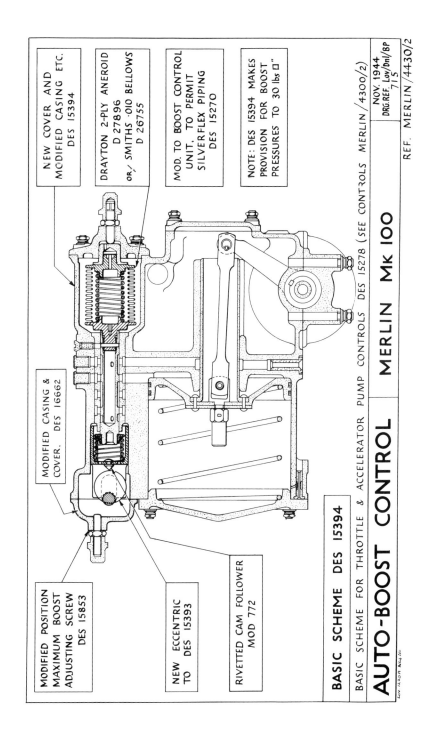

NEW COVER AND MODIFIED CASING ETC. DES 15394

DRAYTON 2-PLY ANEROID D 27896 OR, SMITHS ·010 BELLOWS D 26755

MOD. TO BOOST CONTROL UNIT, TO PERMIT SILVERFLEX PIPING DES 15270

NOTE: DES 15394 MAKES PROVISION FOR BOOST PRESSURES TO 30 lbs ☐"

MODIFIED CASING & COVER. DES 16662

MODIFIED POSITION MAXIMUM BOOST ADJUSTING SCREW DES 15853

NEW ECCENTRIC TO DES 15393

RIVETTED CAM FOLLOWER MOD 772

BASIC SCHEME DES 15394

BASIC SCHEME FOR THROTTLE & ACCELERATOR PUMP CONTROLS DES 15278 (SEE CONTROLS MERLIN/4300/2)

AUTO-BOOST CONTROL | MERLIN Mk 100

NOV. 1944
DRG. REF. Lov/bn//8P
715

REF. MERLIN/4430/2

Lov 14304 Aug bio

ILLUSTRATED NEW FEATURES
OF THE
MERLIN MK. 130
PORT ENGINE, D.H. HORNET

PROVISION FOR AUX. GEARBOX DRIVE.

DE-AERATOR MOUNTED ON REAR S/C. CASING
SEE ILLUSTRATION REF. 4364/2

S/C. INTAKE
DOWN-DRAFT S/C. INTAKE ELBOW. MODIFIED ACCEL. PUMP & INJECTION NOZZLE.
SEE ILLUSTRATION REF. 4400/6

NEW COOLANT PUMP MOUNTED ON I/C. PUMP BRACKET
SEE ILLUSTRATION REF. 4110/5

OIL PIPES & INLET CONN. TO PRESSURE OIL PUMP TO SUIT D.H. HORNET INSTALLATION.

MODIFIED CONNECTIONS TO S.U. INJECTION PUMP
SEE SUPOLDER ISSUE No 4 Jan/NAP

MODIFIED SUPERCHARGER OPERATING MECH. WITH INTEGRAL PNEUMATIC RAM
SEE ILLUSTRATION REF. 4402/2

ENGINE CONTROLS TO SUIT D.H. HORNET INSTALLATION
SEE ILLUSTRATION REF. 4300/5

FEATURES NOT SPECIFIED TO BE AS MERLIN Mk 100 SERIES

LOCKHEED U/C PUMP DRIVEN OFF R.A.E. FACING AT -791 E.S.
DES 17207

FLEXIBLE AIR SCREW OIL PIPES.

REVERSED COOLANT FLOW WITH COOLANT CONNS. TO SUIT D.H. HORNET INSTALLATION.

ALTERED POSITION OF PRIMING PICK-UP CONNS.
SEE ILLUSTRATION REF. 4364/2

SUPERCHARGER
MERLIN 100 REAR CASING INVERTED. DETAIL CHANGES TO VOLUTE DRAIN AND POS-·ITION OF CHANGE OVER VALVE
SEE ILLUSTRATION REF. 4400/6

GENERATOR MOUNTED ON SPECIAL ADAPTOR ON RELIEF VALVE HOUSING
SEE ILLUSTRATION REF. 4140/2

FULLY STRENGTHENED COFFMAN TYPE C./CASE

·420 RATIO REDUCTION GEAR RIGHT HAND ROTATION.
SEE ILLUSTRATION REF. 4240/9

DEC. 1944
Dng. Ref: Lov/Dml/B.P.
719

MERLIN KEY/18.

PRINCIPAL NEW FEATURES OF THE MERLIN Mk130 (R.H. TRACTOR) 'PORT' ENGINE FOR D.H. HORNET AIRCRAFT

118

BLANKING COVER ON
WHEELCASE REPLACES
COOLANT PUMP DRIVE
HOUSING ETC.

BASIC SCHEME DES.15735

BLANKING COVER INCORPORATING LOWER
BALL BEARING HOUSING FOR LOWER VERTICAL DRIVE

MERLIN MK.130

NOV. 1944

DRG.REF: Low/Dn/LS.721

MERLIN/4070/6

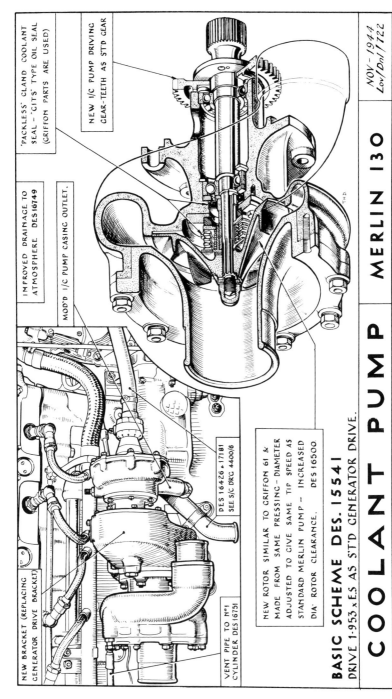

'PACKLESS' GLAND COOLANT
SEAL – "CITS" TYPE OIL SEAL
(GRIFFON PARTS ARE USED)

NEW I/C PUMP DRIVING
GEAR-TEETH AS ST'D GEAR

IMPROVED DRAINAGE TO
ATMOSPHERE DES16749

MOD'D I/C PUMP CASING OUTLET.

NEW BRACKET (REPLACING
GENERATOR DRIVE BRACKET)

VENT PIPE TO N°1
CYLINDER DES 16751

DES 16426 & 17181
SEE S/C DRG 4400/6

NEW ROTOR SIMILAR TO GRIFFON 61 &
MADE FROM SAME PRESSING – DIAMETER
ADJUSTED TO GIVE SAME TIP SPEED AS
STANDARD MERLIN PUMP – INCREASED
DIA° ROTOR CLEARANCE.   DES 16500

BASIC SCHEME DES. 15541

DRIVE 1·953 × E.S AS ST'D GENERATOR DRIVE.

COOLANT PUMP | MERLIN 130

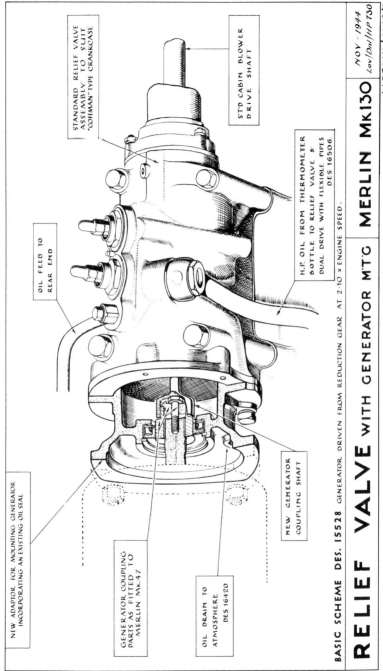

STANDARD RELIEF VALVE ASSEMBLY TO SUIT "COFFMAN" TYPE CRANKCASE

STD CABIN BLOWER DRIVE SHAFT

H.P. OIL FROM THERMOMETER BOTTLE TO RELIEF VALVE & DUAL DRIVE WITH FLEXIBLE PIPES DES 165D6.

OIL FEED TO REAR END

NEW ADAPTOR FOR MOUNTING GENERATOR INCORPORATING AN EXISTING OIL SEAL

GENERATOR COUPLING PARTS AS FITTED TO MERLIN Mk.47

OIL DRAIN TO ATMOSPHERE DES 16420

NEW GENERATOR COUPLING SHAFT

BASIC SCHEME DES. 15528 GENERATOR DRIVEN FROM REDUCTION GEAR AT 2.10 × ENGINE SPEED.

RELIEF VALVE WITH GENERATOR MT'G MERLIN Mk130

NOV - 1944
Lov/Dnl/HP730

MERLIN/4140/2

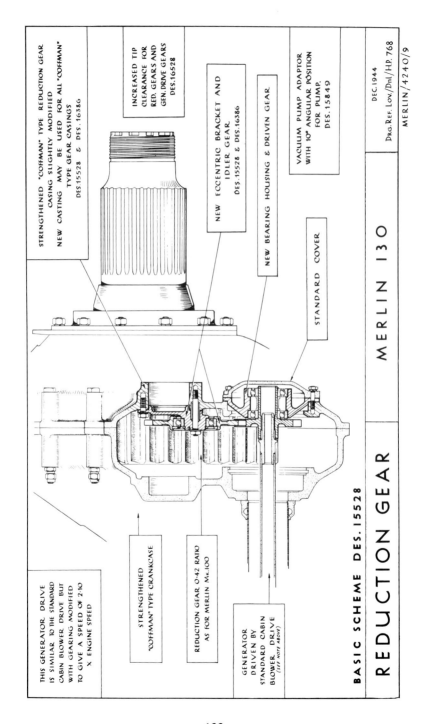

THIS GENERATOR DRIVE IS SIMILAR TO THE STANDARD CABIN BLOWER DRIVE BUT WITH GEARING MODIFIED TO GIVE A SPEED OF 2·10 X ENGINE SPEED

STRENGTHENED "COFFMAN" TYPE REDUCTION GEAR CASING SLIGHTLY MODIFIED NEW CASTING MAY BE USED FOR ALL "COFFMAN" TYPE GEAR CASINGS DES.15528 & DES.16386

INCREASED TIP CLEARANCE FOR RED. GEARS AND GEN. DRIVE GEARS DES.16528

NEW ECCENTRIC BRACKET AND IDLER GEAR DES.15528 & DES.16386

NEW BEARING HOUSING & DRIVEN GEAR

VACUUM PUMP ADAPTOR WITH 10° ANGULAR POSITION FOR PUMP. DES.15849

STANDARD COVER

STRENGTHENED "COFFMAN" TYPE CRANKCASE

REDUCTION GEAR 0·42 RATIO AS FOR MERLIN Mk.100

GENERATOR DRIVEN BY STANDARD CABIN BLOWER DRIVE (SEE NOTE ABOVE)

MERLIN 130

BASIC SCHEME DES.15528

REDUCTION GEAR

DEC.1944

DRG. REF. Lov/Dnl/H.P. 768

MERLIN/4240/9

122

COOLANT CONNS. TO AND FROM INTERCOOLER RAD.
DES.15785 SHORTENED CONNS. DES.16268

COOLANT PIPE I/C PUMP TO S/C DES.16426

NEW POSITION OF DRAIN COCK DES.16966

FRONT COOLANT OUTLET CONNS. TO DES.16729

OIL FEED TO I/C PUMP DRIVE DES.16355

MOD. PIPE. HEADER TANK TO I/C DES.17181

COOLANT CONNS.   MERLIN 130

NOV. 1944
DRG.REF.LOV./DNL./CIR.733

MERLIN/4270/1

DETAIL OF THROTTLE
TORQUE SPRING BALANCE
DES. 16911

INCREASED SET OF
THROTTLE LEVER
DES. 16511

GENERAL ARRANGEMENT
(PORT SIDE OF ENGINE)

SIMPLIFIED FORGING OF PILOT'S LEVER
DES. 16052

PILOT'S PICK-UP CONTROL

NEW INDICATOR
PLATE
DES. 17283

A/BOOST CONTROL UNIT AS M.100
BUT MOD. CONTROL SHAFTS, MOUNTED
REVERSED ON CAST PLATFORM ON
INTAKE ELBOW. —— DES. 15723

BASIC SCHEME DES. 15723

CONTROLS   MERLIN 130

NOV. 1944
DRG. REF. Lov./Dnl./C.I.R. 724

MERLIN/4300/3

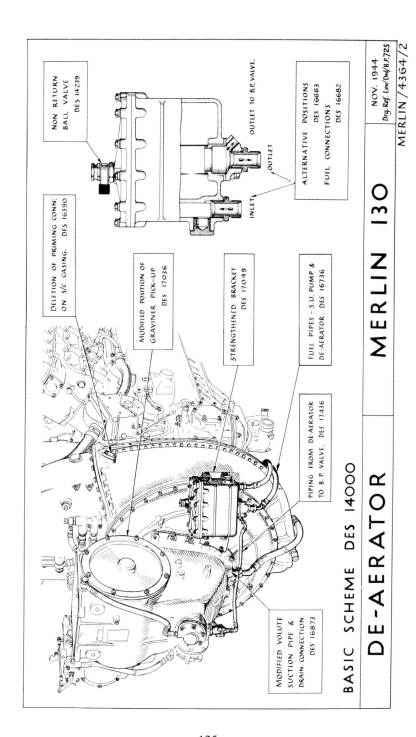

NON RETURN BALL VALVE
DES 14239

DELETION OF PRIMING CONN:
ON S/c CASING. DES 16390

OUTLET TO B.P. VALVE.

OUTLET

INLET

ALTERNATIVE POSITIONS
DES 16663
FUEL CONNECTIONS
DES 16682

MODIFIED POSITION OF
GRAVINER PICK-UP
DES 17036

STRENGTHENED BRACKET
DES 17049

FUEL PIPES - S.U. PUMP &
DE-AERATOR. DES 16736

PIPING FROM DE-AERATOR
TO B.P. VALVE. DES 17436

MODIFIED VOLUTE
SUCTION PIPE &
DRAIN CONNECTION
DES 16873

BASIC SCHEME DES 14000

DE-AERATOR

MERLIN 130

NOV. 1944.
Drg. Ref. Low/Dwl/B.P.725

MERLIN/4364/2

125

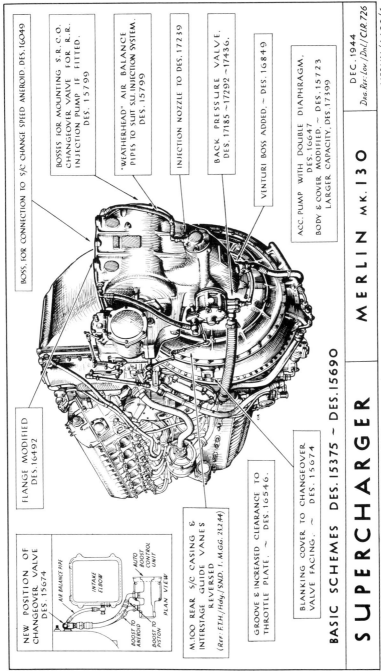

BOSS, FOR CONNECTION TO S/C CHANGE SPEED ANEROID. DES. 16049

FLANGE MODIFIED DES. 16492

BOSSES FOR MOUNTING S. R. C. O. CHANGEOVER VALVE FOR R.R. INJECTION PUMP IF FITTED. DES. 15799

"WEATHERHEAD" AIR BALANCE PIPES TO SUIT S.U. INJECTION SYSTEM. DES. 15799

INJECTION NOZZLE TO DES. 17239

BACK PRESSURE VALVE. DES. 17185 ~ 17292 ~ 17436.

VENTURI BOSS ADDED. ~ DES. 16849

ACC. PUMP WITH DOUBLE DIAPHRAGM. DES. 16647
BODY & COVER MODIFIED. ~ DES. 15723 LARGER CAPACITY. DES. 17399

NEW POSITION OF CHANGEOVER VALVE DES. 15674

AIR BALANCE PIPE

INTAKE ELBOW

AUTO BOOST CONTROL UNIT

PLAN VIEW

BOOST TO ANEROID

BOOST TO PISTON

M.100 REAR S/C CASING & INTERSTAGE GUIDE VANES REVERSED
(Ref: E.T.H./Hdy./SND. 1. M.GG. 23.2.44)

BLANKING COVER TO CHANGEOVER VALVE FACING. ~ DES. 15674

GROOVE & INCREASED CLEARANCE TO THROTTLE PLATE. ~ DES. 16546.

BASIC SCHEMES DES. 15375 ~ DES. 15690

SUPERCHARGER

MERLIN MK. 130

DEC. 1944
Dng. Ref: Lov / Dn./ CIR. 726

MERLIN/4400/6

126

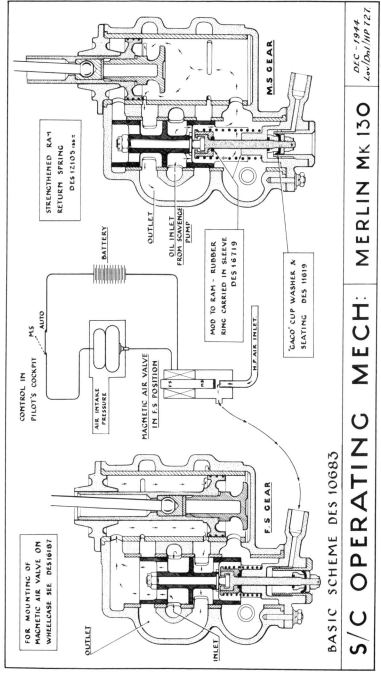

S/C OPERATING MECH: MERLIN MK 130

MERLIN /4402/3

DEC – 1944
Lov/Dnl/HP 727.

BASIC SCHEME DES 10683

M.S GEAR

F.S GEAR

STRENGTHENED RAM
RETURN SPRING
DES 12105 ISSᵈ

BATTERY

CONTROL IN
PILOT'S COCKPIT
M.S
AUTO

AIR INTAKE
PRESSURE

MAGNETIC AIR VALVE
IN F.S POSITION

F.S
M.S

H.P. AIR INLET.

MOD TO RAM – RUBBER
RING CARRIED IN SLEEVE
DES 16719

'GACO' CUP WASHER &
SEATING DES 11619

OUTLET

OIL INLET
FROM SCAVENGE
PUMP

FOR MOUNTING OF
MAGNETIC AIR VALVE ON
WHEELCASE SEE DES 16187

OUTLET

INLET

127

ILLUSTRATED NEW FEATURES
OF THE
MERLIN MK. 131
STARBOARD ENGINE, D.H. HORNET

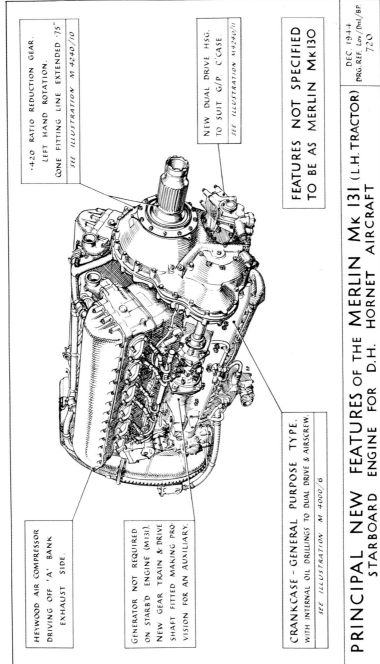

·420 RATIO REDUCTION GEAR.
LEFT HAND ROTATION.
CONE FITTING LINE EXTENDED ·75″
SEE ILLUSTRATION M 4240/10

NEW DUAL DRIVE HSG.
TO SUIT G/P. C'CASE.
SEE ILLUSTRATION M 4240/11

FEATURES NOT SPECIFIED
TO BE AS MERLIN Mk I3O

HEYWOOD AIR COMPRESSOR
DRIVING OFF 'A' BANK
EXHAUST SIDE.

GENERATOR NOT REQUIRED
ON STARB'D ENGINE (M131).
NEW GEAR TRAIN & DRIVE
SHAFT FITTED MAKING PRO-
VISION FOR AN AUXILIARY.
SEE ILLUSTRATION M 4000/6

CRANKCASE - GENERAL PURPOSE TYPE.
WITH INTERNAL OIL DRILLINGS TO DUAL DRIVE & AIRSCREW.
SEE ILLUSTRATION M 4000/6

PRINCIPAL NEW FEATURES OF THE MERLIN Mk 131 (L.H. TRACTOR)
STARBOARD ENGINE FOR D.H. HORNET AIRCRAFT

DEC. 1944
DRG. REF. Lav/Dnl/BP.
720

MERLIN KEY /19

THESE GALLERY PIPES AND DRILLINGS ARE UNUSED WITH DUAL DRIVE MOUNTED ON REDUCTION GEAR.

MAIN PRESSURE OIL
AIRSCREW OIL - INNER
AIRSCREW OIL OUTER

MAIN HIGH PRESSURE OIL FEED FROM PUMP. THERMOMETER HOUSING FACING UNALTERED.

OIL DRILLINGS & GALLERY PIPES REQUIRED WITH C.S.U MOUNTED ON DUAL DRIVE DES 16476 MOD° BRACKETS ETC DES 17000

SPILL FROM RELIEF VALVE

BREATHER CONN. SPIGOTTED & SEALING RING GROOVE DES 17029

¼ PROP SHAFT

OIL TO INNER A'S SHAFT
OIL TO OUTER A'S SHAFT

OIL RETURN FROM C.S.U

H.P. OIL FEED TO C.S.U.

BASIC SCHEMES DES 16190 & 16161

CRANKCASE GENERAL PURPOSE MERLIN MK 131

NOV - 1944
Lov/Dnl/HP 729

MERLIN/4000/6

131

FOR OIL FLOW IN C.P. C'CASE SEE MERLIN/4000/6

FOR OIL FLOW IN DUAL DRIVE SEE MERLIN/4240/11

MOD⁰ OIL JET FOR PINION & IDLER DES 17314

GEAR TOOTH PARTICULARS DES 15846 INCREASED TIP CLEARANCES DES 17183

OIL FEED TO REDUCTION GEAR JETS DES 17430

PIPE TO RELIEF VALVE AND SINGLE BANJO BOLT. DES 17087

MODᴺ TO O'D OF PINION DRIVING SHAFT ; TO FACILITATE GRINDING OF SPLINES DES 16773 ₁₅₅2

MODᴺ TO BORE OF REVERSING IDLER TO PREVENT INCORRECT ASSEMBLY DES 17192

JAN-1945
Lov/Dnl/HP.731

MERLIN/4240/10

BASIC SCHEME DES 15849 REVERSED ROTATION - 422 RATIO.

REDUCTION GEAR | MERLIN Mk131

BLANKED OFF BY FACE OF C.S.U.

OIL TO INNER A'S SHAFT BY-PASSES C.S.U

RETURN OIL FROM C.S.U.

OIL INLET TO C.S.U

OIL FROM C.S.U TO INNER A'S

RETURN OIL FROM C.S.U TO OUTER AIRSCREW SHAFT

OIL TO OUTER A'SCREW SHAFT

CRANKSHAFT FRONT FEED

DUAL DRIVE H'S'G DES 15849

H.P OIL FEED

CHANGEOVER PAD

OIL FLOW IN DUAL DRIVE H'S'G FOR ROTOL ELECTRIC & D.H. HYDROMATIC C.S.U

OIL FLOW IN DUAL DRIVE H'S'G FOR ROTOL HYDRAULIC & D.H BRACKET & DOUBLE ACTING C.S.U

H.P OIL FEED

MERLIN 131

DUAL DRIVE MERLIN

NOV-1944
Lov/Dnl/HP 732
MERLIN/4240/11

MERLIN POWER DEVELOPMENTS

The Merlin started life in 1933 as the PV12 delivering 790 hp with a boost of 2¼lb/sq.in. By the outbreak of the war in 1939, Merlin IIs and IIIs, in service, were producing just over 1000 hp with a boost of 6¼lb/sq.in. on 87 octane fuel. The needs of war accelerated development and by its end, with basically the same crankshaft, but improved pistons, connecting rods, cylinders and valve gear, the Merlin was developing over 2000 hp with 25lb/sq.in. boost and, in development, had cleared for flight test 2340 hp with 30lb/sq.in. boost and completed an endurance test at 2620 hp with 36lb/sq.in. boost.

The 'secret' of this success was boost pressure; the power being proportional to it, but to enable this, supercharger development (both pressure rise and efficiency to reduce its power requirements), fuel development (to enable the use of the higher boost pressures without 'knocking') and of course, mechanical development, (to withstand the increased loads, pressures and temperatures) were all required.

The team responsible for this achievement was led by Cyril Lovesey (Lov), Arthur Rubbra (Rbr) and Stanley Hooker (SGH) until the gas turbine took him away in early 1943. Geoff Wilde (GLW) then assumed his role after working with SGH from 1938 onwards.

Lov again used the talents of Tony Dunwell to 'sell' this development and we have included in this section some of Lov/Dnl's work.

The first step in the improvement in operational service came in the summer of 1940 when '100 octane' fuel (100/130 grade) became available in the squadrons, allowing the boost on the Merlin III to be increased from 6¼ lb/sq.in. to 12 lb/sq.in., raising the power from 1030 hp to 1300 hp but at a lower altitude. Two speed supercharging was introduced to allow boosts to be increased at altitude whilst maintaining them at lower levels, to avoid excessive charge temperatures. Its first application was on the Merlin X where the lower charge temperature and reduced supercharger horsepower gave increased powers at the lower altitudes.

The next step, the improved central entry supercharger, the first significant piece of work contributed by SGH, first saw application in the Merlin XX and 45 series of engines, and with similar levels of boost raised the power levels to around 1400 hp.

The two stage supercharger, essentially a Vulture unit added to the front of the existing Merlin unit, first appeared in squadron service on the Merlin 61 with boost pressures of 15 lb/sq.in. and a horsepower in excess of 1500 hp. Increases in boost pressure to 18lb/sq.in. took the Merlin 63, 64, 72 and 73 to 1680 hp.

Further compressor development, increasing the first stage diameter from 11.5 in to 12.0 in and the replacement of the S.U. carburettor with the Bendix unit, reducing the pressure loss into the compressor, resulted in two families of the engine, the 'high altitude' rating (RM11SM - Merlin 70, 71, 76 and 77) and the 'low altitude' rating (RM10SM - Merlin 65 and 66). Power was optimised to suit the altitude requirements by choice of supercharger gear ratios; lower ratios for lower altitudes and higher ratios for higher altitudes. Maximum power of both these ratings were raised to around 1700 to 1750 hp but the main benefit of the improvement in compressor performance was used to increase the full throttle altitudes by approximately 2000 ft.

The 100 series Merlin introduced further compressor improvements and reduction in inlet pressure losses, by the introduction of the overhung rotor and single point fuel injection and again the benefits were used to raise full-throttle altitude by up to 4000 ft with virtually no increase in power. The 100 series engine, in a similar way to the basic two stage engine, had two ratings for low altitude operation (RM14SM) and high altitude operation (initially in development RM15SM but RM16SM for production).

The Merlin 130s introduced the downdraught carburettor aimed mainly at reducing the frontal area. The 134/135s saw the introduction of the Corliss throttle to reduce the torque effects on the spindle but also provided a cleaner inlet to the supercharger.

Improvements in fuel to 115/150 grade allowed boost pressure to go beyond 20 lb/sq. in. to 25 lb/sq. in. for both 'low altitude' ratings of the 60 and 100 series engine (i.e. Merlin 66 and Merlin 130/131s) and produce powers in excess of 2000 hp.

The final development (RM17SM) came from improvements in the supercharger, with the rotor diameters being increased to 12.7 in. for the first stage and 10.7 in. for the second stage, together with a change in valve timing with a longer duration exhaust cam with increased overlap. This engine was type tested at 2200 hp with 30 lb/sq.in. and cleared for flight at 2340 hp, however with the end of the war, it never went into squadron service and therefore never received a Ministry Mark number.

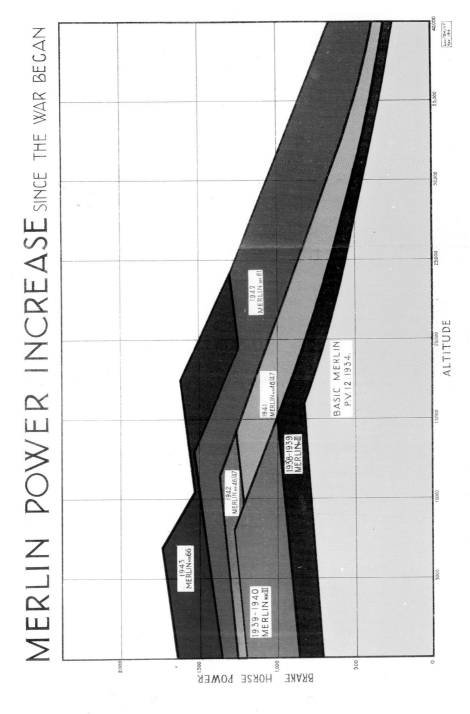

MERLIN POWER INCREASE SINCE THE WAR BEGAN

BRAKE HORSE POWER

ALTITUDE

1943 MERLIN MK.66

1939-1940 MERLIN MK.III

1942 MERLIN MK.46/47

1936-1939 MERLIN MK.III

1941 MERLIN MK.46/47

1942 MERLIN MK.61

BASIC MERLIN P.V.12. 1934.

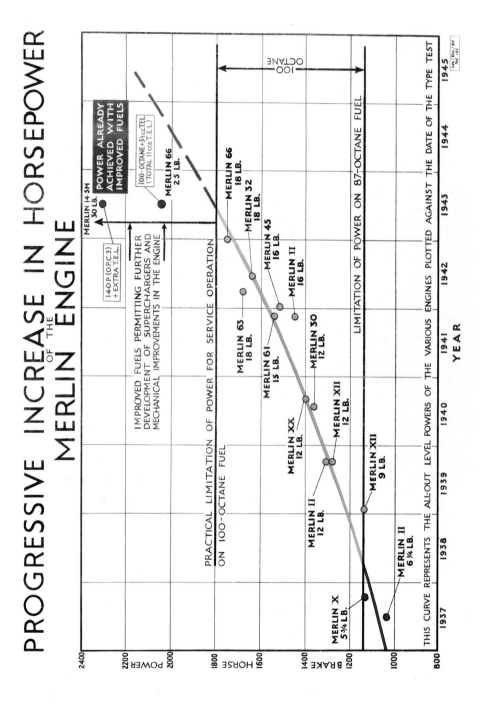

PROGRESSIVE INCREASE IN HORSEPOWER
OF THE MERLIN ENGINE

100 OCTANE

POWER ALREADY ACHIEVED WITH IMPROVED FUELS

MERLIN 14 SM
30 LB.

140P.(O.P.C.3)
+ EXTRA T.E.L.

100-OCTANE+5½ccTEL)
(TOTAL 11 ccs T.E.L.)

MERLIN 66
25 LB.

MERLIN 66
18 LB.

MERLIN 32
18 LB.

MERLIN 45
16 LB.

MERLIN II
16 LB.

MERLIN 63
18 LB.

MERLIN 61
15 LB.

MERLIN 30
12 LB.

MERLIN XX
12 LB.

MERLIN XII
12 LB.

MERLIN II
12 LB.

MERLIN XII
9 LB.

MERLIN II
6¾ LB.

MERLIN X
5¾ LB.

IMPROVED FUELS PERMITTING FURTHER
DEVELOPMENT OF SUPERCHARGERS AND
MECHANICAL IMPROVEMENTS IN THE ENGINE

PRACTICAL LIMITATION OF POWER FOR SERVICE OPERATION

PRACTICAL LIMITATION OF POWER
ON 100-OCTANE FUEL

LIMITATION OF POWER ON 87-OCTANE FUEL

THIS CURVE REPRESENTS THE ALL-OUT LEVEL POWERS OF THE VARIOUS ENGINES PLOTTED AGAINST THE DATE OF THE TYPE TEST

BRAKE HORSE POWER

2400
2200
2000
1800
1600
1400
1200
1000
800

YEAR

1937 1938 1939 1940 1941 1942 1943 1944 1945

138

MERLIN DEVELOPMENT
TO MEET SPECIAL HIGH ALTITUDE CONDITIONS

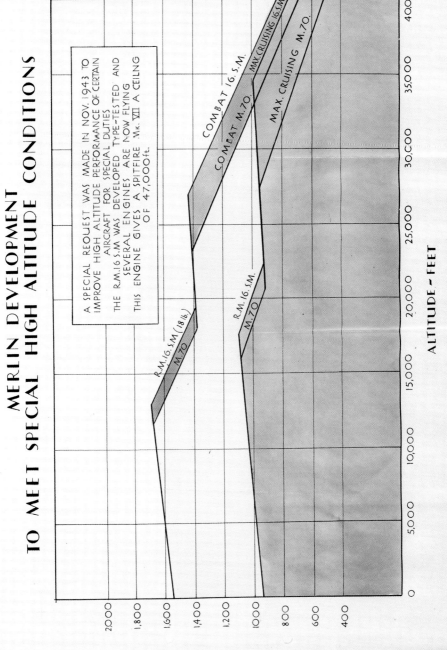

A SPECIAL REQUEST WAS MADE IN NOV. 1943 TO IMPROVE HIGH ALTITUDE PERFORMANCE OF CERTAIN AIRCRAFT FOR SPECIAL DUTIES

THE R.M. 16 S.M WAS DEVELOPED TYPE-TESTED AND SEVERAL ENGINES ARE NOW FLYING

THIS ENGINE GIVES A SPITFIRE Mk. VII A CEILING OF 47,000 ft.

ALTITUDE ~ FEET

COMBAT 16. S.M.
COMBAT M. 70.
MAX. CRUISING 16. S.M
MAX. CRUISING M. 70.
R.M. 16. S.M. (18 lb.)
M. 70
R.M. 16. S.M.
M. 70

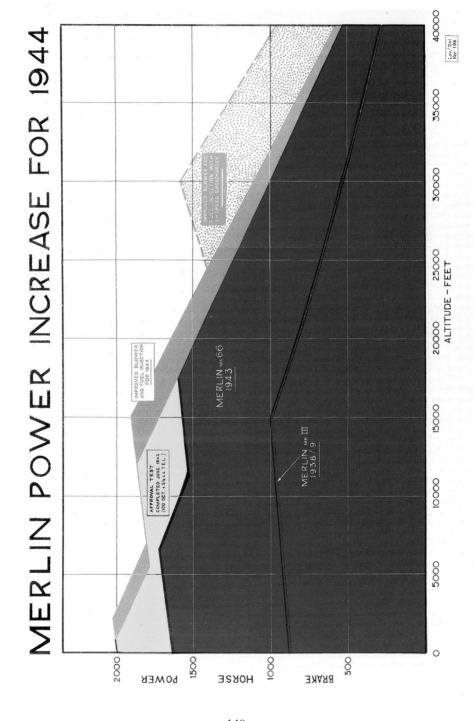

140

Rolls-Royce Merlin 100 Series
RM14SM Rating

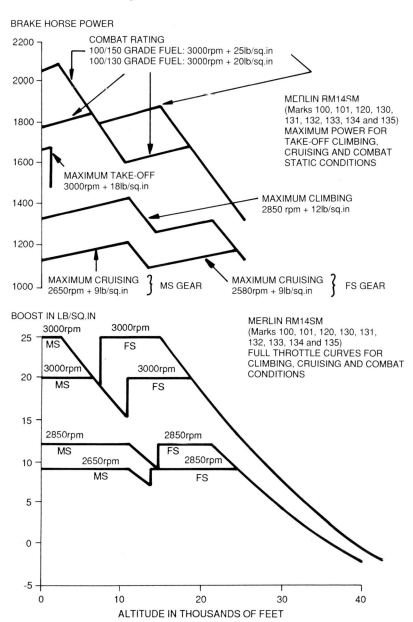

BRAKE HORSE POWER

COMBAT RATING
100/150 GRADE FUEL: 3000rpm + 25lb/sq.in
100/130 GRADE FUEL: 3000rpm + 20lb/sq.in

MERLIN RM14SM
(Marks 100, 101, 120, 130,
131, 132, 133, 134 and 135)
MAXIMUM POWER FOR
TAKE-OFF CLIMBING,
CRUISING AND COMBAT
STATIC CONDITIONS

MAXIMUM TAKE-OFF
3000rpm + 18lb/sq.in

MAXIMUM CLIMBING
2850 rpm + 12lb/sq.in

MAXIMUM CRUISING } MS GEAR
2650rpm + 9lb/sq.in

MAXIMUM CRUISING } FS GEAR
2580rpm + 9lb/sq.in

BOOST IN LB/SQ.IN

MERLIN RM14SM
(Marks 100, 101, 120, 130, 131,
132, 133, 134 and 135)
FULL THROTTLE CURVES FOR
CLIMBING, CRUISING AND COMBAT
CONDITIONS

3000rpm MS
3000rpm FS
3000rpm MS
3000rpm FS
2850rpm MS
2850rpm FS
2650rpm MS
2850rpm FS

ALTITUDE IN THOUSANDS OF FEET

141

Rolls-Royce Merlin 100 Series
RM16SM Rating

BRAKE HORSE POWER

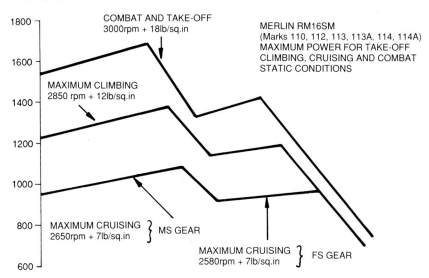

MERLIN RM16SM
(Marks 110, 112, 113, 113A, 114, 114A)
MAXIMUM POWER FOR TAKE-OFF
CLIMBING, CRUISING AND COMBAT
STATIC CONDITIONS

COMBAT AND TAKE-OFF
3000rpm + 18lb/sq.in

MAXIMUM CLIMBING
2850 rpm + 12lb/sq.in

MAXIMUM CRUISING
2650rpm + 7lb/sq.in } MS GEAR

MAXIMUM CRUISING
2580rpm + 7lb/sq.in } FS GEAR

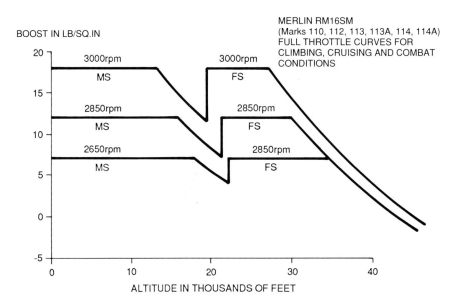

BOOST IN LB/SQ.IN

MERLIN RM16SM
(Marks 110, 112, 113, 113A, 114, 114A)
FULL THROTTLE CURVES FOR
CLIMBING, CRUISING AND COMBAT
CONDITIONS

3000rpm   3000rpm
MS   FS

2850rpm   2850rpm
MS   FS

2650rpm   2850rpm
MS   FS

ALTITUDE IN THOUSANDS OF FEET

Merlin 100 Series
Comparison of combat ratings

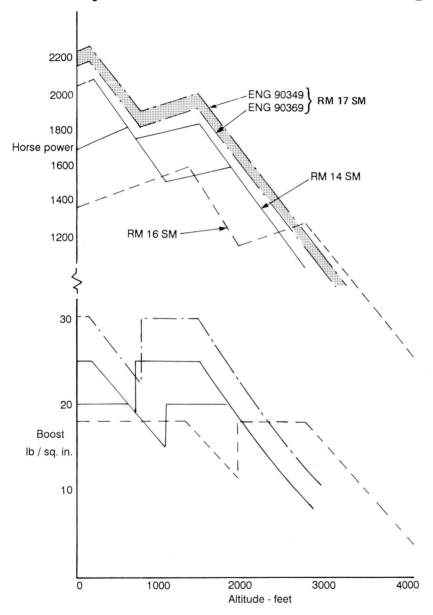

DEVELOPMENT HISTORY OF THE MERLIN 100 SERIES

After the Merlin 66 had passed it's type test in Nov/Dec of 1942 on Engine 82861, attention was turned to the next stage of development. In the Spring of 1943, Merlin 66s were running up to 25 lb/sq.in. boost and with S.U. fuel injection pumps providing the experience for the specification of the 100 series to be issued in April. A batch of twenty five 100 series engines was ordered by the Controller of Research and Development (CRD at the Ministry) to be built by 'Experimental' at Derby and the engines (odd numbers from 90347 to 90395) were delivered between August 1943 and August 1944.

In July 1943, before these engines were delivered, a significant milestone was reached when a 'Merlin 66' No. 82861 ran to 3300 rpm and 30 lb/sq.in. boost and delivered 2380 hp during a 15 min. 'Sprint' test.

The first two 100 series engines, 90347 and 90349 carried out 114 hr type tests in August 1943, before 90347 attempted the 100 hr overload test, which was curtailed at 77½ hr by engine failure. The overload test was successfully completed by engine 90353 (see the next chapter for further details) in November; following strip, inspection and reassembly this engine was then installed in Mustang FX 858 at Hucknall for endurance flying.

Ten engines from the batch were supplied to de Havilland for use in the five Mosquito PR 32 aircraft. The prototype PR32, MM328, was flown to Hucknall in June 1944, and the remaining four, NS586 to NS589 were delivered from Hatfield in September and October.

In July and August of 1944, two further RM16SM engines (No. 120401 and 120403) were built and tested by Experimental to support the 100 series Development programme. The summer of 1944 had also seen the arrival of the 'Schneiderised' Merlin 130, with six engines (No. 190359 to 190369) being built and tested in May and June with a further four 131 engines (No. 190370 to 190376 - left hand tractors) following in August through October. Two engines (190359 and 190370) were allocated to Development whilst the other eight were delivered to de Havilland for use in the two Hornet prototype aircraft (RR915 and RR919). The first prototype RR915 flew in July 1944 with two 130s fitted (both right hand tractors) and it was not until October after some 50 hours flying that the first Merlin 131 (No. 190372) was fitted. The installations were such that the propeller blades passed upwards adjacent to the fuselage. Although take-off swing was improved, tip/fuselage interference effects resulted in paddle blades being fitted, and finally the installations were swopped over to give downward tip motion adjacent to the fuselage, before acceptable characteristics were obtained. These Experimental Merlin 131s were also used to support the early production aircraft RX210 through RX213 until the production 131s were delivered in June 1945.

Problems encountered early in the flight programme with throttle response, were traced to the aerodynamic loads on the butterfly valves; 'Corliss' throttles were fitted and tested on Hornet RX212 at Hucknall in early 1946 and were incorporated onto Production in the Merlin 134/135s in the later part of the same year.

During 1944 engines 90349 and 90369 were converted to the RM17SM rating. In April/May, 90349 whilst attempting a 10 hr flight approval test, failed after 6 hours due to blade rod bearing breakup which was of an earlier development standard. The test was successfully completed in August on Engine 90369, which although to the same standard as 90349, produced about 100 hp less. In spite of this, the engine was inadvertently run to a higher power than intended during the 30 min. high power run in MS gear; 2340 hp being recorded at 3000 rpm and 30 lb/sq.in. boost. The RM17SM rating was type tested later that year at 2200 hp. Engine 90369 was flown at Hucknall in Mustang FX 858 and details are given in 'Rolls-Royce and the Mustang' (No. 9 in our series).

Towards the end of the year a Merlin 66 No. 185547 was converted to the RM17SM rating and on 12 December ran to 2620 hp at 3150 rpm with 36 lb/sq.in. boost and water injection, completing a 15 min. 'Sprint' run. This was the highest power to be recorded on a Merlin during the war years.

MERLIN DEVELOPMENT WITH IMPROVED FUELS.

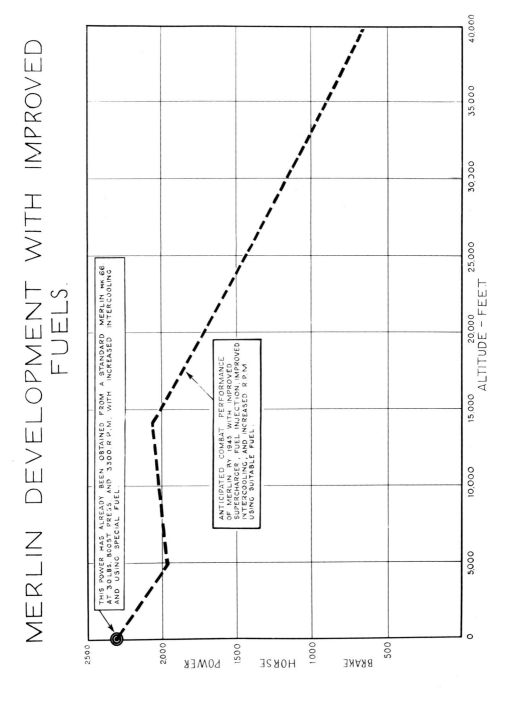

THIS POWER HAS ALREADY BEEN OBTAINED FROM A STANDARD MERLIN MK 66 AT 30 LBS. BOOST PRESS AND 3300 R.P.M. WITH INCREASED INTERCOOLING AND USING SPECIAL FUEL.

ANTICIPATED COMBAT PERFORMANCE OF MERLIN BY 1945 WITH IMPROVED SUPERCHARGER, FUEL INJECTION, IMPROVED INTERCOOLING, AND INCREASED R.P.M. USING SUITABLE FUEL.

BRAKE HORSE POWER

ALTITUDE – FEET

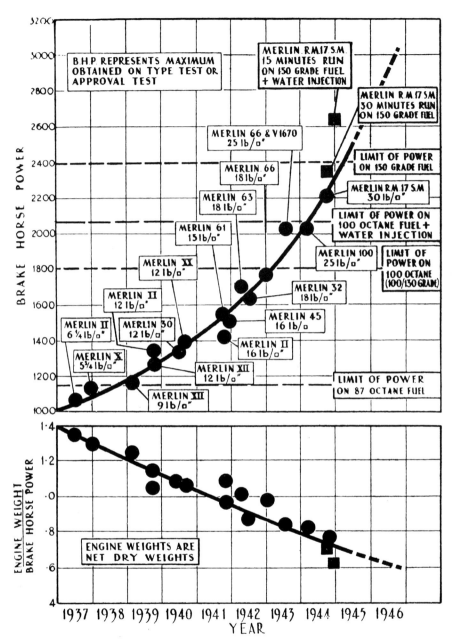

Rolls-Royce Merlin—progressive improvement in power
and weight power ratio with time

148

POWER CURVES PRIOR TO HIGH POWER RUNS

Engine No.		82861			90369		185547	
Mark No.		66			RM17SM		RM17SM	
Date		15.7.43			18.8.44		12.12.44	
R.P.M.	3000	3150	3300	2850	3000	3150	3000	3150
Boost Hg.abs.	76.97	84.07	91.37	85.27	91.0	91.0	95.46	101.46
Back Press. Hg.abs.	31.07	31.07	31.07	30.97	30.97	31.07	30.51	30.91
Air Temp. °C	22	22	22	22	22	22	10	10
Charge Temp. °C	40	44	50	78	84	88	56	47
I/C Water Temp 'IN' °C	12.5	12.5	12.5	55	55	55	5	5
Coolant Outlet Temp. °C	75	80	84	1.02	105	105	80	37
Oil Inlet Temp. °C	40	45	50	90	90	90	52	51
Fuel Pts/Hr.	1315	1410	1650	1450	1525	1670	1575	1880
Water Pts/Hr.	–	–	–	–	–	–	0	300
Specific Cons. pts/BHP/hr.	.648	.66	.720	.745	.725	.792	658	.829
B.H.P. obsd.	2025	2140	2298	1942	2120	2110	2394	2632
B.H.P. (fully corrected)	2095	2220	2380	2015	2139	2129	2385	2620
NOTES	Curve prior to 15 min. "Sprint" test FUEL O.P.C.3 +16.8 cc T.E.L.			Curve prior to 10 hr. approval test. Throttled to +30 lb/sq.in. FUEL RDE/F/290			Curve prior to 15 min. "Sprint" test. FUEL RDE/F/290 plus water as a bi-fuel	

MERLIN 100 SERIES DEVELOPMENT ENGINES

ENGINE NO.	MARK NO./RATING	ALLOCATED TO	DESPATCHED TO
90347		DEVELOPMENT	
90349	100	DEVELOPMENT	RR GLASGOW – 31-10-45
90351			RR HUCKNALL – 6-12-43 – SPITFIRE JL106
90353		DEVELOPMENT	
90355	100(SP)		RR HUCKNALL – LANCASTER DV199
90357	100(SP)		RR HUCKNALL – LANCASTER DV199
90359	100(SP)		RR HUCKNALL – LANCASTER DV199
90361		?	
90363		?	
90365	RM16SM		DE HAVILLAND, HATFIELD – 21-2-44
90367	RM16SM		DE HAVILLAND, HATFIELD – 22-2-44 – MOSQUITO MM 328
90369		DEVELOPMENT	
90371	100(SP)		RR HUCKNALL – LANCASTER DV199
90373	RM16SM		DE HAVILLAND, HATFIELD – 12-4-44 – MOSQUITO MM 328
90375			RR HUCKNALL – SPITFIRE JL106
90377	100(S)	DEVELOPMENT	GLOUCESTER? – 26-10-44
90379	RM16SM		DE HAVILLAND, HATFIELD – 19-4-44
90381	114(S)/RM16SM		DE HAVILLAND, HATFIELD – 3-6-44
90383	112(S)/RM16SM		SUPERMARINE, HIGH POST – 3-6-44
90385			DE HAVILLAND, HATFIELD – 3-6-44
90387	RM14SM	PACKARD?	NOW IN SMITHSONIAN, WASHINGTON DC
90389	114		DE HAVILLAND, HATFIELD – 21-7-44
90391	114		DE HAVILLAND, HATFIELD – 18-7-44
90393	114		DE HAVILLAND, HATFIELD – 10-8-44
90395	114		DE HAVILLAND, HATFIELD – 7-8-44

MERLIN 100 SERIES DEVELOPMENT ENGINES

ENGINE NO.	MARK NO./RATING	ALLOCATED TO	DESPATCHED TO
120401	RM16SM	DEVELOPMENT	
120403	114/RM16SM	DEVELOPMENT	
190359	130/RM14SM	DEVELOPMENT	
190361	130/RM14SM		DE HAVILLAND, HATFIELD – 21-5-44 – HORNET RR915
190363	130/RM14SM		DE HAVILLAND, HATFIELD – 21-5-44 – HORNET RR915
190365	130/RM14SM		DE HAVILLAND, HATFIELD – 16-9-44
190367	130/RM14SM		DE HAVILLAND, HATFIELD – 19-8-44
190369	130/RM14SM		DE HAVILLAND, HATFIELD – 20-11-44
190370	131/RM14SM	DEVELOPMENT	
190372	131/RM14SM		DE HAVILLAND, HATFIELD – 21-9-44
190374	131/RM14SM		DE HAVILLAND, HATFIELD – 13-11-44
190376	131/RM14SM		DE HAVILLAND, HATFIELD – 11-11-44

ROLLS-ROYCE MERLIN 100 SERIES
DEVELOPMENT HISTORY

Columns (years / months): 1943 — JUN JUL AUG SEP OCT NOV DEC · 1944 — JAN FEB MAR APR MAY JUN JUL AUG SEP OCT NOV DEC

ENGINE NUMBER

- **SUPPLY OF DEVELOPMENT ENGINES**
- **90347 – 90395** — 100 SERIES – 25 ENGINES
- **120401 – 120403** — RM16SM-2
- **190359 – 190369** — 130 SERIES-6
- **190370 – 190376** — 131 SERIES-4
- **SIGNIFICANT TESTS**
- **82861** — M66 AT 2340 HP · T.T. – TYPE TEST – 114 HRS
- **90347** — T.T. RM14SM · O.L. (77 HRS) · O.L. – OVERLOAD TEST – 100 HRS · T.T. RM14SM/251b
- **90349** — T.T. RM14SM
- **90353** — RM14SM O.L.
- **90369** — RM16SM T.T. · RM17SM F.A.T. (6 HRS) · RM17SM · RM17SM T.T.
- **90377** — T.T. · F.A.T.
- **185547** — T.T. RM14SM · RM17SM AT 2620 HP · RM17SM T.T.
- **190359** — RM14SM T.T.

F.A.T. – FLIGHT APPROVAL TEST – 10 HRS

152

ROLLS–ROYCE MERLIN 100 SERIES
DEVELOPMENT FLYING AT HUCKNALL

ENGINE NUMBER	1943	1944	1945	Flying record (chronological, as labelled)
90351				RM14SM; SPITFIRE JL106, 50.5 HRS; RM16SM (113); 43.1 HRS
90353				RM14SM; MUSTANG FX858, 50.8 HRS; MUSTANG FX901; FX858 6.5 HRS
90355				LANCASTER DV199, 170.7 HRS
90357				LANCASTER DV199, 170.7 HRS
90359				LANCASTER DV199, 170.7 HRS
90367				MOSQUITO MM328, 10.8 HRS
90369				RM17SM; MUSTANG FX858, 0.5 HRS; MUSTANG FX858; RM17SM; 8.7 HRS
90371				LANCASTER DV199, 170.7 HRS
90373				MOSQUITO MM328, 10.8 HRS; RM16SM (112)
90375				SPITFIRE JL106; RM16SM (112); 135.1 HRS; 45.25 HRS
111493				SPITFIRE MD176; RM16SM (74/112) 26.75 HRS; 11.75 HRS; SPITFIRE MD176
88253				RM16SM (61/112); 1.1 HRS; SPITFIRE MD176

1943: DEC — 1944: JAN FEB MAR APR MAY JUN JUL AUG SEP OCT NOV DEC — 1945: JAN FEB MAR APR MAY JUN JUL

DEVELOPMENT OF MECHANICAL FEATURES TO TAKE CARE OF INCREASED POWER

It was realised that the normal service type test, whilst being good proof of the reliability of the engine at its particular rating, was not sufficiently arduous to reveal quickly the mechanical limitations which were likely to arise in the next stage of development. Therefore a basic overload test was adopted, first with the object of finding the mechanical weaknesses of existing Marks of engines and secondly with the object of arriving at a specification which would satisfactorily pass the test. The overload test consisted of 100 hours endurance at 3000 rpm and 18lb boost pressure which was the highest combat condition for engines in service at that time, and the object was to complete the test without any adjustment or replacements.

The overload testing was started with a standard production Merlin 66 but failed due to cracks in the crankcase after 27 hours. To make sure that this was not a rogue failure the crankcase was replaced but again failed after about the same running time. In order not to hold up progress, testing was continued by replacing the failed parts and the endurance running proceeded on the serviceable features until such time as modified designs were available for testing.

A total of 1000 hours running under these conditions were completed on various experimental engines before eventually achieving the target of 100 hours on an engine which completed the test in a very satisfactory condition. By this time the engine had reached what was called the Mark 100 specification. The test was completed successfully in six days and the engine ran perfectly throughout. There were no involuntary stops and the usual routine maintenance was omitted entirely, even the valve, plugs and tappets being untouched throughout the whole of the test. The only incident was the change of one magneto after 50 hours, due to a defect revealed by the single ignition check. The condition of the engine when stripped was excellent. The engine was then rebuilt without any replacement parts being fitted and was endurance tested in Mustang FX 858 at Hucknall.

The illustrations show some of the problems encountered during this overload development and the final conditions of the engine which passed the test.

Development continued through 1944 to address the remaining problems and some of the solutions are shown in the Dunwell drawings published in the winter of 1944/45 and shown in the earlier chapter.

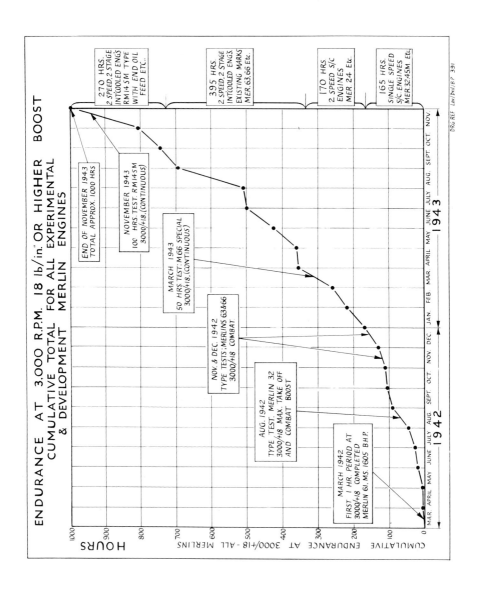

ENDURANCE AT 3,000 R.P.M. 18 lb/in." OR HIGHER BOOST
CUMULATIVE TOTAL FOR ALL EXPERIMENTAL
& DEVELOPMENT MERLIN ENGINES

270 HRS.
2 SPEED, 2 STAGE
INTOOLED ENGS
RM14 SM TYPE
WITH END OIL
FEED ETC..

395 HRS
2 SPEED, 2 STAGE
INTOOLED ENGS
EXISTING MARKS
MER. 63, 66 Etc.

170 HRS
2 SPEED S/C
ENGINES
MER 24 Etc.

165 HRS.
SINGLE SPEED
S/C ENGINES
MER 32, 45M Etc.

END OF NOVEMBER 1943
TOTAL APPROX. 1000 HRS

NOVEMBER 1943
100 HRS. TEST. RM14SM
3000/+18. (CONTINUOUS)

MARCH 1943
50 HRS TEST: M66 SPECIAL
3000/+18. (CONTINUOUS)

NOV. & DEC. 1942
TYPE TESTS: MERLINS 63, 66
3000/+18 COMBAT

AUG. 1942
TYPE TEST. MERLIN 32
3000/+18 MAX. TAKE OFF
AND COMBAT BOOST

MARCH 1942
FIRST 1 HR. PERIOD AT
3000/+18 COMPLETED
MERLIN 61. MS. 1605 B.H.P.

DRG. REF Low/Dril/B.P. 391

1943

1942

HOURS

1000
900
800
700
600
500
400
300
200
100
0

CUMULATIVE ENDURANCE AT 3000/+18 - ALL MERLINS

MAR. APRIL MAY JUNE JULY AUG. SEPT OCT NOV DEC JAN FEB MAR APRIL MAY JUNE JULY AUG. SEPT OCT NOV

RELIABILITY WITH CONTINUOUS OPERATING CONDITIONS OF 18 lbs/in² BOOST & 3000 RPM

LIMITATIONS OF STANDARD COMPONENTS.	EXPERIENCE OF IMPROVED PERFORMANCE.
MAIN BEARING PANEL FAILURE. FAILURE OF CRANKCASES IN MAIN BEARING PANELS AFTER ENDURANCE SHOWN TESTS(1)-(2) 27¼ Hrs (3) 2½ " (4) (6) 79¼ " **STANDARD CRANKCASE**	NO FAILURES CRANKCASES SERVICEABLE AFTER - TESTS (7) (13) 148 Hrs (CONVENTIONAL OIL FEED) TESTS (14) 87½ Hrs (15) 100 - (END FEED OIL FEED) DEEPER BOSSES AND STUDS. STRENGTHENED MAIN BEARING CAPS **STRENGTHENED CRANKCASE**
OIL GROOVES AND HOLES TO FEED CONN ROD BEARINGS. REJECTED FOR EXCESSIVE WEAR AFTER - TESTS (4)-(6) 79¼ Hrs **STANDARD MAIN BEARINGS**	OIL GROOVES AND HOLES DELETED WITH END FEED. NO WEAR MEASUREABLE AFTER - TESTS (14) 87½ Hrs (15) 100 " **STRENGTHENED MAIN BEARINGS.**
CRACK. BROKEN THROUGH No 4 CRANKPIN BY FATIGUE CRACK STARTING FROM THREADS TEST (14) 87½ Hrs **END OIL FEED CRANKSFT FIRST DESIGN**	THREADS DELETED & METAL ADDED. NO FAILURE. CRANKSHAFT SERVICEABLE AFTER - TEST (15) 100 Hrs **END OIL FEED CRANKSFT STRENGTHENED**
3 NARROW GAS RINGS 2 SCRAPER RINGS. SOME RING GUMMING AFTER - TEST (1) 9½ Hrs (2) 7½ " (3) 21½ " (5) 20 " (6) 50 " **STANDARD PISTONS**	DEEP TOP LAND 3 STANDARD GAS RINGS TOP SCRAPER RING DELETED. NO RING GUMMING PISTONS IN GOOD CONDITION AFTER - TEST (14) 87½ Hrs (15) 100 Hrs **DEEP TOP LAND PISTONS**

157

DEVELOPMENT OF RM14SM MECHANICAL FEATURES
FOR IMPROVED RELIABILITY AND INCREASED RATING

RELIABILITY WITH CONTINUOUS OPERATING CONDITIONS OF 18 lbs/in² BOOST & 3000 R.P.M.

LIMITATIONS OF STANDARD COMPONENTS	EXPERIENCE OF IMPROVED PERFORMANCE

LIMITATIONS OF STANDARD COMPONENTS

CYLINDER HEADS CRACKED FROM NOS 3 & 4 TOP CORE PLUGS AFTER—
TESTS (1)-(3)   48¼ HRS
          (A' & B')
TESTS (4)-(6)   79¼ "
          ('B' ONLY)

STANDARD CYLINDER HEADS

CYLINDER SKIRTS CRACKED FROM SIDE STUD BOSS OR COOLANT CONNECTIONS AFTER—
TESTS (1)-(3) A SIDE   48¼ HRS.
        (4)-(6) A SIDE   79¼ "
        (1)-(6) B SIDE   128 "
        (7)-(9) A SIDE   72 "
        (9)-(13) A SIDE   76 "
        (13)('A'&B)   100 "

STANDARD CYLINDER SKIRTS

FRONT PINION ROLLER BEARING FAILED AFTER—
TESTS (1)-(2)   27¾ HRS
        (3)-(5)   50 ¾ "

STANDARD REDUCTION GEAR PINION BEARINGS

SUPERCHARGER BALL BEARING FAILED AFTER—
TEST (1)   19 ½ HRS
       (2)-(4)   38½ "

STANDARD S/C BALL BEARING

EXPERIENCE OF IMPROVED PERFORMANCE

NOS 3 & 4 CORE PLUGS DELETED

NO FAILURES
HEAD SERVICEABLE AFTER
TESTS (7)-(13)   148 HRS
        (15)   100 "

NOS 3 & 4 CORE PLUGS DELETED CYLINDER HEADS

STRENGTHENED SIDE GIRDER

NO FAILURES
SKIRTS SERVICEABLE AFTER
TEST (14)   87½ HRS
    "   (FURTHER TESTS IN HAND)

STRENGTHENED CYLINDER SKIRTS

DIVIDED COOLANT CONNECTIONS

LARGER ROLLERS

PINION BEARINGS IN EXCELLENT CONDITION
AFTER —
TEST (15)   100 HRS

INCREASED CAPACITY REDUCTION GEAR PINION BEARINGS

INTEGRAL MOUNTING FLANGE PERMITTING LARGER BALLS

SUPERCHARGER BALL BEARING IN GOOD CONDITION AFTER —
TEST (14)   87½ HRS
       (15)   100 HRS

INTEGRAL FLANGE S/C BALL BEARING

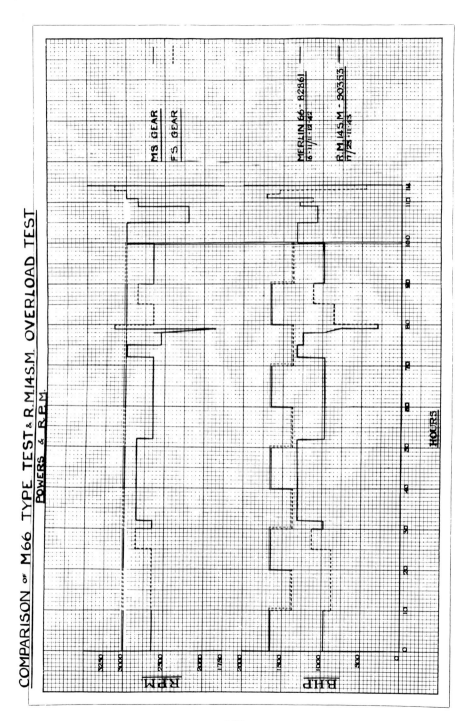

COMPARISON OF M66 TYPE TEST & R.M.14S.M. OVERLOAD TEST
POWERS & R.P.M.

MS GEAR
FS GEAR

MERLIN 66 - 82261
6.11/II 12.42

R.M.14S.M - 903553
17/25 11.43

HOURS

159

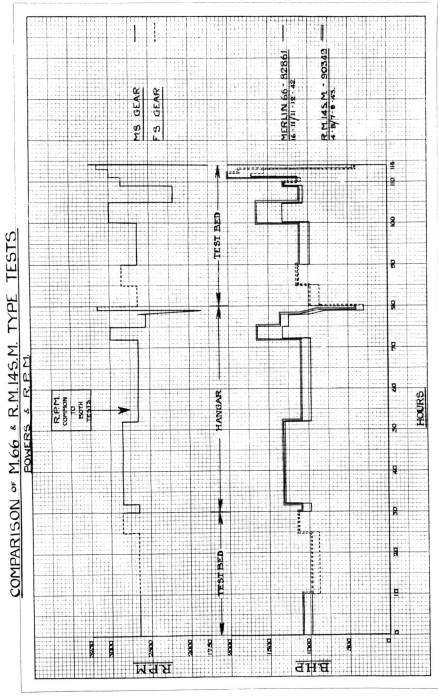

COMPARISON OF M.66 & R.M.14S.M. TYPE TESTS
POWERS & R.P.M.

160

DEVELOPMENT OF RM14SM MECHANICAL FEATURES FOR IMPROVED RELIABILITY & INCREASED RATING.

RECORD OF ENDURANCE TESTS AT 18 lb./□" BOOST 3000 R.P.M. CONTINUOUSLY

TEST NO.	ENGINE NO.	TEST DATE.	PRINCIPAL ITEMS OR CHANGES IN MECHANICAL SPECIFICATION.	ENDUR. COMP. IN TEST.	RESULT OF TEST.	ENGINE TOTALS 3000 / +18 ENDURANCE.	ENGINE TOTALS RUNNING TIME.
1	M.82425	5th to 6th December 1942	New engine, standard Merlin 66 specification except for 0.42:1 reduction gear and cabin blower drive	19¾	Supercharger ball bearing failed	19¾	35¼
2	"	12th to 14th December 1942	New standard S/C. ball bearing fitted.	7½	Crankcase failed; 2 and 3 main bearing panels broken away. Reduction gear pinion front bearing roller track breaking up.	27¼	53¼
3	"	16th to 17th January 1943	New standard crankcase and reduction gear fitted.	21½	Crankcase failed; Nos. 2, 5 and 6 main bearing panels broken away. Both cylinder heads cracked from Nos. 3 and 4 top core plugs; "A" skirt cracked.	48¾	91
4	"	20th to 22nd February 1943	New standard crankcase, main bearings and cylinder heads and skirt fitted. S.U. Mark II fuel injection pump fitted.	9½	Supercharger ball bearing failed.	58¼	110
5	"	2nd to 3rd March 1943	Fischer ball bearing fitted to S/C; same external dimensions as standard but larger balls and thinner races.	19¾	Reduction gear pinion front roller bearing failed.	78	141
6	"	15th to 19th March 1943	Reduction gear with 0.471:1 ratio 26½° pressure angle teeth and increased capacity pinion bearings with larger rollers fitted. (This was an intermediate design scheme; the gear unit was not new, having completed a Merlin 32 Type Dev. Test, total running time 175 hours).	50	50 HOUR ENDURANCE COMPLETED. Crankcase cracked in No.4 panel; main bearings badly worn. Crankshaft cracked in Nos. 1 and 6 crankpins, ascribed to effect of running in broken crankcases in Tests 2 and 3. "A" cylinder head cracked from No.4 top core plug and both skirts cracked. S/C. main drive ball bearing broken up and S/C. ball bearing rejected owing to loss of running clearance.	128	199
7	"	7th to 10th May 1943	Rebuilt with parts of new design; strengthened crankcase cylinder heads with Nos. 3 and 4 top core plugs deleted; increased capacity S/C. main drive ball bearing in stiffened panel; S/C. ball bearing with integral mounting flange and larger balls.	30¼	S.U. fuel injection pump seized. S/C. main drive ball bearing and housing in bad condition.	158¼	252¼

Continued....

RECORD OF ENDURANCE TESTS AT 18 lb/□" BOOST 3000 R.P.M. CONTINUOUSLY

TEST NO.	ENGINE NO.	TEST DATE.	PRINCIPAL ITEMS OR CHANGES IN MECHANICAL SPECIFICATION.	ENDUR. CONT. IN TEST.	RESULT OF TEST.	ENGINE TOTALS 3000 / +18 ENDURANCE.	ENGINE TOTALS RUNNING TIME.
8	M.82425	19th to 21st May 1943	New S.U. pump and S/C. main drive ball bearing fitted.	3¾	Piston failed, ascribed to breakage of scraper ring. "A" cylinder skirt cracked.	162½	263½
9	"	7th to 10th June 1943	Deep top land, 3 standard gas ring pistons; half sets of Wilcox Rich (U.S.A.) and R.R. welded head exhaust valves fitted.	37½	M.S. clutch withdrawal race failed (this was the original part fitted to the engine). No.1 main bearing seized, damaging crankshaft. "A" cylinder skirt cracked.	200	308¾
10	"	3rd to 5th August 1943	Standard replacement parts fitted. Also new increased capacity main drive ball bearing with improved flanged housing; increased load valve springs.	30¾	Exhaust valve (Wilcox - Rich) cracked and burned through head.	230¾	350¼
11	"	9th August 1943	Replacement exhaust valve fitted.	2½	S/C. ball bearing failed; this bearing had run on another engine and was found to be one from a batch which had been incorrectly heat treated.	233¼	355
12	"	15th August 1943	Special experimental S/C. ball bearing fitted (increased number of smaller balls).	5¼	Upper vertical drive shaft ball bearing failed.	238½	368
13	"	20th to 25th August 1943	Standard replacement parts fitted.	37½	Special S/C. ball bearing failed owing to breakage of cage.	276	413
14	M.90347	26/8 to 8th September 1943	This was one of three experimental engines built to a basic mechanical specification issued in April 1943 and intended to be the basic engine for the R.M.14.S.M. Features are :- Strengthened crankcase and main bearing caps, increased flow end oil feed crankshaft and main bearings; deep top land 3 standard gas ring pistons; improved connecting rods; strengthened cylinder blocks; increased load valve springs; improved coolant pump; S.U. Mark II fuel injection pump; 0.95" spring drive shaft with 8½° stops, 3-plate starfish M.S. clutch; Merlin 66 S/C. with integral flange bearing; 0.471 reduction gear with improved tooth form, increased capacity pinion bearings with larger rollers and strengthened casing. N.B. This engine had completed 114 hours Type Development Test at R.M.14.S.M. rating before this endurance test at 3000/+18; total running time 166½ hours including 10 hours at 3000/+18 or higher boost.	77½	Crankshaft broken through No. 4 crankpin by fatigue crack starting from threads for screwed plug in oilway. (Crankshaft redesigned at this point, threads deleted and metal added to webs). Engine generally in good condition.	87½	272

Continued....

RECORD OF ENDURANCE TESTS AT 18 lb/□" BOOST 3,000 R.P.M. CONTINUOUSLY

TEST No.	ENGINE No.	TEST DATE.	PRINCIPAL ITEMS OR CHANGES IN MECHANICAL SPECIFICATION.	ENDUR. COMP. IN hrs.	RESULT OF TEST.	ENGINE TOTALS 3000/+18 ENDURANCE.	RUNNING TIME.
15	M.90353	18th to 23rd November 1943	R.M.14.S.M. Basic Engine. This was first of 24 preliminary R.M.14.S.M. engines built to a basic specification issued in June 1943, for Flight trials here and in the U.S.A. The specification was similar to that for Merlin 90347 above except for :- Standard cylinders but with Nos. 3 and 4 top core plugs deleted (fully strengthened cylinders not available at time of build); standard Merlin 61 type connecting rods; 0.42 reduction gear with increased capacity pinion bearings and strengthened casing; "starfish" F.S. clutch. N.B. This engine was fitted with strengthened end oil feed crankshaft and overhung first stage supercharger.	100	100 Hour Test at 3000/+18 Completed. The engine was found to be in remarkably good condition when stripped for examination; the main bearings in particular were virtually as new. The strengthened crankshaft, overhung first stage S/C. and other special features were all satisfactory. The principal defects noted were :- 3 "A" bank exhaust rockers and cams scuffed, "A" and "B" bank cylinder skirts cracked from side stud shoulder and one stud broken; local break up of the surface of one tooth of the reduction gear pinion and damage to one F.S. clutch, apparently due to sludging up.	100	122

MERLINS 82425, 90347 AND 90353 ONLY :-

TOTAL ENDURANCE AT 3000/+18 = 463½ HOURS.

TOTAL RUNNING TIME = 807 HOURS.

163

R.M.14 S.M. No.90353, November 18th - 23rd 1943.

Total Running Time - 122 Hours.

Unit	Condition	Action Necessary
CRANKCASE Strengthened	Excellent; no cracks whatever.	None
MAIN BEARINGS L.T.1. Lead plated; no grooves or holes, for end oil feed lubrication.	Excellent; unmarked and wear negligible.	None
CRANKSHAFT Strengthened, end oil feed type.	Excellent; no wear.	None
PISTONS Double girder, Circular, 3 Standard gas ring.	Excellent; all rings free.	None
CONNECTING-RODS Standard Merlin 66.	Excellent.	None
VALVES Standard.	Good; all seats intact but heads of 6 exhaust valves somewhat corroded.	None; but welded Nimonic head exhaust valves are being developed.
CYLINDERS Liners, standard. Heads, standard except for deleted Nos. 3 & 4 top core plugs. Skirts, standard.	Excellent; no ridging and wear extremely small. Good; one end side stud broken, and few small cracks from stud tubes. Poor; both cracked at side stud shoulder in No.5. cylinder.	None))) Strengthened) cylinders are) instructed for) Merlin) Production.)
SPRING DRIVE To suit end oil feed.	Excellent.	None

164

Unit	Condition	Action Necessary
WHEELCASE & GEARS Standard, modified to suit end oil feed.	Fair; bevel gears driving camshafts worn.	None
CAM DRIVES	Fair; bevel gears worn.	None
COOLANT PUMP Sealed bearing type.	Excellent.	None
MAGNETO DRIVE Standard, with rubber couplings.	Excellent.	None
CAM AND ROCKER MECHANISM Standard, but with increased oil feed.	Poor; 3 exhaust rockers scuffed on "A" side, damaging camshaft.	Increased base circle cams (with slow closing toe) already instructed for development.
REDUCTION GEAR Increased capacity pinion roller bearings.	Excellent.	None
Gear wheel, Standard Merlin 61 type 0.42:1 ratio.	Excellent.	))
Pinion.	Poor; surface of one tooth breaking up at root and 3 teeth with surface cracks.	) Tests with modi-) fied tip relief to) be made.) Investigation of) surface cracking) proceeding.
Strengthened casing.	Excellent.	None
DUAL DRIVE To suit end oil feed.	Excellent.	None
FUEL PUMP S.U. Mark II.	Excellent.	None
SUPERCHARGER Overhung 1st stage type, integral flange ball bearing.	Excellent.	Mods to reduce oil leakage into induction system to prevent sludging of inter-cooler. 1. Reduced clearance on metering plug for intermediate bearing. 2. Schemes to reduce oil leakage at front ball bearing.

Unit	Condition	Action Necessary
SUPERCHARGER DRIVES 3 Plate M.S. clutch.	Excellent.	Oil drains to be added in view of F.S. clutch trouble.
Central web M.S. gear, 5.79:1 ratio.	Poor; teeth show surface cracking.	This is a general problem now under investigation.
M.S. layshaft thrust bearing, standard.	Poor; ball-race pitted.	Under investigation.
Starfish F.S. clutches, 7.06:1 ratio.	Poor; one burned out, due to sludging up.	Oil drain holes added and oil flow reduced.
Main drive, increased capacity ball bearing, with flanged housing and stiffened panel.	Fair; bearing in excellent condition but loose in housing which was worn; this led to increased backlash at main bevel which accounts for excessive wear of bevels driving camshaft.	Improved fit of bearing in housing.
INTERCOOLER Cast box type.	Excellent mechanically; air flow restricted by sludging during test.	Mods. to S/C. oil seals to prevent entry of oil to induction system.

PRODUCTION HISTORY OF THE MERLIN 100 SERIES

The Merlin 100 Series was manufactured by Rolls-Royce at both Hillington in Glasgow and Derby and by Packard in the USA. The brief histories are shown on the following two charts. It is worth noting that the majority of the 102s and 104s, which as mentioned earlier, never established themselves in service, and were finally converted to Merlin 134/135s in 1948/1949, the 102s first being converted to Civil 600s to support the development of the Civil 620 Series, before they in turn were finally converted.

Production started at Packard in April 1945 with the - 9 and - 9A for the Mustang. Although the records indicate that they also made some Merlin 300s and 301s for installation in Avro Lincolns B Mk 2 and Canadian built de Havilland Mosquitoes FB Mk 24 respectively, there is no evidence that they were ever installed.

ROLLS–ROYCE MERLIN 100 SERIES
PRODUCTION HISTORY

PACKARD MERLIN 100 SERIES
PRODUCTION HISTORY

1945

MARK NUMBER	APRIL	MAY	JUNE	JULY	AUGUST	SEPTEMBER	OCTOBER
V-1650-9	▨	▨	▨	▨	▨	▨	
V-1650-9A	▨	▨	▨	▨	▨	▨	
300		▨	▨	▨			
301			▨	▨			
V-1650-23					▨	▨	
V-1650-25						▨	▨

Notes –

The exact numbers produced are not known, but in the closing months of the war,
Packard were producing about 1000 engines per month including the basic V1650-1 and -7
Production was brought to a sudden halt soon after VJ day, and its not clear exactly when
100 series production halted but from the available evidence it was around the end of September.

ROLLS–ROYCE MERLIN 600 SERIES
PRODUCTION HISTORY

MARK NUMBER LISTING FOR PROJECTED AND PRODUCTION ENGINES

This section reproduces facsimiles of the listings of Mark Numbers maintained and issued by the Rolls-Royce Production Drawing Office and the Service Department by way of their Aero Service Bulletins.

For completeness, we have included charts for all the two stage engines including the civil variants.

ROLLS-ROYCE PISTON ENGINES
LIST OF M.O.S. MARK NUMBERS

Production Mark Number	M.O.S. Experimental Nomenclature	Supercharger		Propeller Reduction Gear Ratio	Max. Boost P.S.I.		Aircraft Type	Remarks
		Rotor Dia.	Gear Ratios		100/130 Grade Fuel	115/150 Grade Fuel		
Merlin 55	RM.5S	10.25	9.09	.477	+16	–	Spitfire	Similar to Merlin 45 but with two-piece cylinder blocks
Merlin 55A	RM.5S	10.25	9.09	.477	+16	–	Spitfire	Similar to Merlin 45 but with lower modification standard
Merlin 55M	RM.5S	9.5	9.09	.477	+18	–	Spitfire	Similar to Merlin 55 but with 'cropped' supercharger rotor
Merlin 55MA	RM.5S	9.5	9.09	.477	+18	–	Spitfire	Similar to Merlin 45M but with two-piece cylinder blocks
Merlin 56	RM.6S	10.85	9.09	.477	+16	–	Spitfire	Similar to Merlin 46 but with diaphragm controlled fuel feed in carburetter
Merlin 60	RM.6SM	11.5/10.1	5.52/8.41	.42	+12	–	Wellington	Basic type, two-speed, two-stage supercharger
Merlin 61	RM.8SM	11.5/10.1	6.39/8.03	.42	+15	–	Spitfire	Similar to Merlin 60 but with different supercharger drive gear ratios and two-piece cylinder blocks
Merlin 62	RM.6SM	11.5/10.1	5.52/8.41	.42	+12	–	Wellington	Similar to Merlin 60 with two-piece cylinder blocks
Merlin 63	RM.8SM	11.5/10.1	6.39/8.03	.477	+18	–	Spitfire	Similar to Merlin 61 but no provision for cabin blower, also with strengthened supercharger drive shaft

ROLLS-ROYCE PISTON ENGINES
LIST OF M.O.S. MARK NUMBERS

Production Mark Number	M.O.S. Experimental Nomenclature	Supercharger Rotor Dia	Supercharger Gear Ratios	Propeller Reduction Gear Ratio	Max. Boost P.S.I. 100/130 Grade Fuel	Max. Boost P.S.I. 115/150 Grade Fuel	Aircraft Type	Remarks
Merlin 63A	RM.8SM	11.5/10.1	6.39/8.03	.477	+18	-	Spitfire	Merlin 63 engine with a Merlin 64 type crankcase but with cabin blower drive parts deleted and blanking covers fitted
Merlin 64	RM.8SM	11.5/10.1	6.39/8.03	.477	+18	-	Spitfire	Similar to Merlin 63 but with drive for cabin blower
Merlin 65	RM.10SM	12.0/10.1	5.79/7.06	.42	+18	-	Mustang (Project only)	Similar to Merlin 63 but fitted with lower supercharger drive ratios, .42 propeller reduction gear and with Bendix carburetter
Merlin 66	RM.10SM	12.0/10.1	5.79/7.06	.477	+18	-	Spitfire	Similar to Merlin 65 but with .477 gear and interconnected controls
Merlin 67	RM.10SM	12.0/10.1	5.79/7.06	.42	+18	-		Similar to Merlin 66 but reversed flow cooling and .42 reduction gear No production
Merlin 68		12.0/10.1	5.802/7.349	.42	+18	-	Lincoln	Packard version of Merlin 85
Merlin 68A		12.0/10.1	5.80/7.34	.42	+18	-	Lincoln	Merlin 68 with charge temperature control scheme

ISSUED BY PRODUCTION DRAWING OFFICE • ROLLS-ROYCE LIMITED DERBY

DATE ISSUED FEBRUARY 1953
Page M.7

ROLLS-ROYCE PISTON ENGINES
LIST OF M.O.S. MARK NUMBERS

Production Mark Number	M.O.S. Experimental Nomenclature	Supercharger		Propeller Reduction Gear Ratio	Max. Boost P.S.I.		Aircraft Type	Remarks
		Rotor Dia.	Gear Ratios		100/130 Grade Fuel	115/150 Grade Fuel		
Merlin 69		12.0/10.1	5.80/7.34	.42	+18	-	Mosquito	Packard version of Merlin 67
Merlin 70	RM.11SM	12.0/10.1	6.39/8.03	.477	+18	-	Spitfire	Similar to Merlin 66 but with higher supercharger drive gear ratios
Merlin 71	RM.11SM	12.0/10.1	6.39/8.03	.477	+18	-		Similar to Merlin 70 but with cabin blower drive. No production
Merlin 72	RM.8SM	11.5/10.1	6.39/8.03	.42	+18	-	Mosquito Welkin	Similar to Merlin 63 but with reversed flow coolant
Merlin 73	RM.8SM	11.5/10.1	6.39/8.03	.42	+18	-	Mosquito	Similar to Merlin 72 but with cabin blower drive
Merlin 76	RM.11SM	12.0/10.1	6.39/8.03	.42	+18	-	Mosquito	Similar to Merlin 72 but with Bendix carburetter
Merlin 77	RM.11SM	12.0/10.1	6.39/8.03	.42	+18	-	Mosquito	Similar to Merlin 76 but with cabin blower drive
Merlin 85	RM.10SM	12.0/10.1	5.79/7.06	.42	+18	-	Lincoln	Generally similar to Merlin 66 but with .42 propeller reduction gear and intercooler header tank integral with intercooler
Merlin 85A	RM.10SM	12.0/10.1	5.79/7.06	.42	+18	-	Lincoln	Similar to Merlin 85 but with improved modification standard

ROLLS-ROYCE PISTON ENGINES
LIST OF M.O.S. MARK NUMBERS

Production Mark Number	M.O.S. Experimental Nomenclature	Supercharger Rotor Dia.	Supercharger Gear Ratios	Propeller Reduction Gear Ratio	Max. Boost P.S.I. 100/130 Grade Fuel	Max. Boost P.S.I. 115/150 Grade Fuel	Aircraft Type	Remarks
Merlin 85B	RM.10SM	12.0/ 10.1	5.79/ 7.06	.42	+18	-	Lincoln	Similar to Merlin 85 with American 4.G.8 constant speed unit
Merlin 86	RM.10SM	12.0/ 10.1	5.79/ 8.03	.42	+18	-	Lincoln	As Merlin 85A for high altitude with anti-surge diffusers
Merlin 90	RM.20SM	11.95	7.0/ 8.15	.42	+18	-	Avro Tudor (Project only)	Similar to Merlin 100 with two-speed single stage supercharger
Merlin 100	RM.14SM	12.0/ 10.1	5.79/ 7.06	.42	+20	-		Basic type with S.U. fuel injection pump No production
Merlin 101	RM.14SM	12.0/ 10.1	5.79/ 7.06	.42	+20	-		Similar to Merlin 100 but with reversed flow cooling No production
Merlin 102	RM.14SM	12.0/ 10.1	5.79/ 7.06	.42	+20	-	Tudor	Similar to Merlin 100 but with strengthened drive for gearbox
Merlin 102A	RM.14SM	12.0/ 10.1	5.79/ 7.06	.42	+20	-	Tudor	Similar to Merlin 102 but incorporating an after-heater
Merlin 104	RM.14SM	12.0/ 10.1	5.79/ 7.06	.42	+20	+25	Mosquito	Similar to Merlin 114 but with lower supercharger drive gear ratio

ISSUED BY PRODUCTION DRAWING OFFICE • ROLLS-ROYCE LIMITED DERBY   DATE ISSUED FEBRUARY 1953

Page M.9

ROLLS-ROYCE PISTON ENGINES
LIST OF M.O.S. MARK NUMBERS

Production Mark Number	M.O.S. Experimental Nomenclature	Supercharger Rotor Dia.	Supercharger Gear Ratios	Propeller Reduction Gear Ratio	Max. Boost P.S.I. 100/130 Grade Fuel	Max. Boost P.S.I. 115/150 Grade Fuel	Aircraft Type	Remarks
Merlin 105	RM.14SM	12.0/10.1	5.79/7.06	.42	+20	–		As Merlin 102 but with 100 H.P. auxiliary gearbox drive No production
Merlin 110	RM.16SM	12.0/10.1	6.39/8.03	.4707	+18	–	Spitfire (Project only)	Basically as Merlin 100 except reduction gear and supercharger ratios
Merlin 112	RM.16SM	12.0/10.1	6.39/8.03	.4707	+18	–	Spitfire (Project only)	As Merlin 110 except for provision for cabin blower
Merlin 113	RM.16SM	12.0/10.1	6.39/8.03	.42	+18	–	Mosquito	Similar to Merlin 100 but with high supercharger gear ratio and reversed flow cooling
Merlin 113A	RM.16SM	12.0/10.1	6.39/8.03	.42	+18	–	Mosquito	Similar to Merlin 113 but with anti-surge supercharger diffusers
Merlin 114	RM.16SM	12.0/10.1	6.39/8.03	.42	+18	–	Mosquito	Similar to Merlin 113 but with cabin supercharger drive
Merlin 114A	RM.16SM	12.0/10.1	6.39/8.03	.42	+18	–	Mosquito	Similar to Merlin 114 but with anti-surge supercharger diffusers
Merlin 120	RM.14SM	12.0/10.1	5.79/7.06	.375	+20	–		Basically as Merlin 100 except for contra-rotating reduction gear No production

Production Mark Number	M.O.S. Experimental Nomenclature	Supercharger		Propeller Reduction Gear Ratio	Max. Boost P.S.I.		Aircraft Type	Remarks
		Rotor Dia.	Gear Ratios		100/130 Grade Fuel	115/150 Grade Fuel		
Merlin 130	RM.14SM	12.0/10.1	5.79/7.06	.42	+20	+25	Hornet	Basic type similar to Merlin 100 series but with down draught intake elbow
Merlin 131	RM.14SM	12.0/10.1	5.79/7.06	.42 L.H.	+20	+25	Hornet	Similar to Merlin 130 but with propeller reduction gear shaft left hand rotation
Merlin 132	RM.14SM	12.0/10.1	5.79/7.06	.42	+20	+25	Hornet	Similar to Merlin 130 but suitable for propeller breaking constant speed unit
Merlin 133	RM.14SM	12.0/10.1	5.79/7.06	.42	+20	+25	Hornet	Similar to Merlin 132 but with propeller reduction gear shaft left hand rotation
Merlin 134	RM.12SM	12.0/10.1	5.79/7.06	.42	+20	+25	Hornet	Similar to Merlin 130 but with corliss throttle and fixed ignitior.
Merlin 135	RM.14SM	12.0/10.1	5.79/7.06	.422	+20	+25	Hornet	Similar to Merlin 134 but with propeller reduction gear shaft left hand rotation
Merlin 140	RM.14SM	12.0/10.1	5.79/7.06	.512	+20	+25	Sturgeon	Similar to Merlin 100 series but contra-rotating reduction gear shunt system of cooling and Coffman starter
Merlin 224	RM.3SM	10.25	8.15/9.49	.42	+18	-	Lancaster	American built Merlin 24

ISSUED BY PRODUCTION DRAWING OFFICE • ROLLS-ROYCE LIMITED DERBY   DATE ISSUED   FEBRUARY 1953

Page M.11

ROLLS-ROYCE PISTON ENGINES
LIST OF M.O.S. MARK NUMBERS

Production Mark Number	M.O.S. Experimental Nomenclature	Supercharger		Propeller Reduction Gear Ratio	Max. Boost P.S.I.		Aircraft Type	Remarks
		Rotor Dia.	Gear Ratios		100/130 Grade Fuel	115/150 Grade Fuel		
Merlin 225	RM.3SM	10.25	8.15/9.49	.42	+18	–	Mosquito	American built Merlin 25
Merlin 266	RM.10SM	12.0/10.1	5.80/7.34	.479	+18	–	Spitfire	American built Merlin 66
Merlin 300	RM.14SM	12.0/10.1	5.80/7.34	.42	+20	–	Lincoln (Project only)	American built Merlin 100
Merlin 301	RM.14SM	12.0/10.1	5.80/7.34	.42	+20	–	Lincoln (Project only)	American built Merlin 101
Merlin V1650-1	RM.3SM	10.25	8.15/9.49	.477	+18	–	Kittyhawk	Corresponds to Merlin 28 (American built Merlin)
Merlin V1650-3		12.0/10.1	6.39/8.095	.479	+18¼	–	Mustang	American built Merlin
Merlin V1650-5		12.0/10.1	6.39/8.095	.479	+18¼	–	King Cobra	American built Merlin No production
Merlin V1650-7	RM.10SM	12.0/10.1	5.80/7.34	.479	+18¼	–	Mustang	Corresponds to Merlin 68 (American built Merlin)
Merlin V1650-9	RM.16SM	12.0/10.1	6.39/8.095	.479	+20	–	Mustang	Water-methanol injection Simmonds power control (American built Merlin)

ISSUED BY PRODUCTION DRAWING OFFICE • ROLLS-ROYCE LIMITED DERBY

DATE ISSUED   FEBRUARY 1953

ROLLS-ROYCE PISTON ENGINES
LIST OF M.O.S. MARK NUMBERS

Production Mark Number	M.O.S. Experimental Nomenclature	Supercharger		Propeller Reduction Gear Ratio	Max. Boost P.S.I.		Aircraft Type	Remarks
		Rotor Dia.	Gear Ratios		100/130 Grade Fuel	115/150 Grade Fuel		
Merlin V1650-9A	RM.16SM	12.0/10.1	6.39/8.095	.479	+20	-	Mustang	Simmonds power control (American built Merlin)
Merlin V1650-11	RM.16SM	12.0/10.1	6.39/8.095	.479	+20	-	Mustang	Water-methanol injection Simmonds power control (American built Merlin)
Merlin V1650-13		12.0/10.1	6.39/8.095	.479	$+18\frac{1}{4}$	-	Mustang	Simmonds power control (American built Merlin)
								No production
Merlin V1650-15		12.0/10.1	6.39/8.095	.479	$+18\frac{1}{4}$	-	Mustang	Simmonds power control (American built Merlin)
								No production
Merlin V1650-17	RM.10SM	12.0/10.1	5.80/7.34	.479	$+18\frac{1}{4}$	-	Mustang	Simmonds power control (American built Merlin)
								No production
Merlin V1650-19		12.0/10.1	5.0 to 8.3	.479	$+19\frac{3}{4}$	-		Supercharger gear ratio variable between limits given (American built Merlin)
								No production
Merlin V1650-21		12.0/10.1	6.39/8.095	.479	+20	-	N.A.A. XP.82	Reversed rotation propeller (American built Merlin)

ISSUED BY PRODUCTION DRAWING OFFICE • ROLLS-ROYCE LIMITED DERBY

DATE ISSUED   FEBRUARY 1953

Page M.13

ROLLS-ROYCE PISTON ENGINES
LIST OF COMMERCIAL MARK NUMBERS

Mark Number	Customer	Supercharger Rotor Dia.	Supercharger Gear Ratios	Propeller Reduction Gear Ratio	Max. Boost lb/sq.inch. 100/130 Grade Fuel	Aircraft Type	Remarks
Merlin 24S	Union of South Africa	10.25	8.15/9.49	.42	+18	–	Pressure water-glycol cooled two-piece cylinder blocks
Merlin T24-4	Trans Canadian Airways	10.25	8.15/9.49	.42	+18	Lancastrian	Similar to Merlin 24 but improved modification standard and incorporating after-heater
Merlin 25-8	Turkish Government	10.25	8.15/9.49	.42	+18	Mosquito	Similar to Merlin 24 but with improved modification standard and reversed flow cooling
Merlin 66-8	Turkish Government	12.0/10.1	5.79/7.06	.477	+18	Spitfire	Two-speed, two-stage supercharger
Merlin 134-5	Argentine Government	12.0/10.1	5.79/7.06	.42	+18	–	Two-speed, two-stage supercharger, with down draught intake elbow Right hand tractor
Merlin 135-5	Argentine Government	12.0/10.1	5.79/7.06	.422	+18	–	Similar to Merlin 134-5 but with left hand tractor
Merlin 150		12.0/10.1	5.79/7.06	.4707	+20	–	Mark number changed to 620
Merlin 151		12.0/10.1	5.79/7.06	.42	+20	–	Mark number changed to 621
Merlin 500	B.O.A.C.	10.25	8.15/9.49	.42	+18	Lancastrian	Pressure water-glycol cooled two-piece cylinder blocks

| Mark Number | Customer | Supercharger | | Propeller Reduction Gear Ratio | Max. Boost lb/sq/inch. 100/130 Grade Fuel | Aircraft Type | Remarks |
		Rotor Dia.	Gear Ratios				
Merlin 504	–	10.25	8.15/9.49	.42	+18	–	As Merlin 502 but incorporating water-methanol equipment for take-off
Merlin 501	–	10.25	8.15/9.49	.42	+18	–	No production
							As Merlin 500 but with gearbox drive and modified intake elbow
Merlin 530	–	11.95	7.0/8.15	.42	+18	–	No production
							Two-speed single stage supercharger and S.U. Injection Pump
Merlin 539	–	9.75	8.588	.422 or .4707	+18	–	No production
							Single speed, single stage supercharger and R.R. fuel injector pump
Merlin 549	–	9.75	8.588	.422	+18	–	No production
							Similar to Merlin 539 but with 6.5 compression ratio pistons
Merlin 600	B.O.A.C.	12.0/10.1	5.79/7.06	.42	+20	Tudor	No production
							Two-speed, two-stage supercharger and incorporating an after-heater

ROLLS-ROYCE PISTON ENGINES
LIST OF COMMERCIAL MARK NUMBERS

Mark Number	Customer	Supercharger		Propeller Reduction Gear Ratio	Max. Boost lb/sq inch. 100/130 Grade Fuel	Aircraft Type	Remarks
		Rotor Dia.	Gear Ratios				
Merlin 604	Argentine Government	12.0/10.1	5.79/7.06	.42	+20	–	Two-speed, two-stage supercharger
Merlin 620	T.C.A.	12.0/10.1	5.79/7.06	.4707	+20	D.C.4	Two-speed, two-stage supercharger and combined intercooler and after-heater Special installation features for D.C.4 aircraft, including American type splines on the propeller shaft converted to inter type
Merlin 621	Basic type	12.0/10.1	5.79/7.06	.42	+20	Tudor	Similar to Merlin 620 but with installation features to suit Tudor aircraft and standard English splines on propeller shaft
Merlin 621-1	Initially operated by B.O.A.C.	12.0/10.1	5.79/7.06	.42	+20	Tudor	As Merlin 621 but with improved modification standard
Merlin 621-2	Initially operated by B.S.A.A.	12.0/10.1	5.79/7.06	.42	+20	Tudor	As Merlin 621 type engines
Merlin 621-5	Flota Aerea Mercante Argentine	12.0/10.1	5.79/7.06	.42	+20	Tudor	As Merlin 621 type engines

ROLLS-ROYCE PISTON ENGINES
LIST OF COMMERCIAL MARK NUMBERS

Mark Number	Customer	Supercharger Rotor Dia.	Supercharger Gear Ratios	Propeller Reduction Gear Ratio	Max. Boost lb/sq.inch. 100/130 Grade Fuel	Aircraft Type	Remarks
Merlin 621-15	Argentine Government	12.0/10.1	5.79/7.06	.42	+20	Lincoln	As Merlin 621 but with installation features for Lincoln aircraft
Merlin 622-10	Trans Canada Airways	12.0/10.1	5.79/7.06	.4707	$+20\frac{1}{2}$	D.C.4	As Merlin 620 but with increased boost for take-off and climb
Merlin 623	Basic type	12.0/10.1	5.79/7.06	.42	$+20\frac{1}{2}$	–	As Merlin 621 but with increased boost for take-off and climb
Merlin 623-2	B.S.A.A.	12.0/10.1	5.79/7.06	.42	$+20\frac{1}{2}$	Tudor	As Merlin 623 type engines
Merlin 624	Basic type	12.0/10.1	5.79/7.06	.42	$+20\frac{1}{2}$	–	As Merlin 622 but with .42 reduction gear ratios
Merlin 624-10	Canadair	12.0/10.1	5.79/7.06	.42	$+20\frac{1}{2}$	–	As Merlin 624 type engines

| Mark Number | Customer | Supercharger | | Propeller Reduction Gear Ratio | Max. Boost lb/sq/inch. 100/130 Grade Fuel | Aircraft Type | Remarks |
		Rotor Dia.	Gear Ratios				
Merlin 625	Basic type	12.0/ 10.1	5.79/ 7.06	.42	+20½	–	Similar to Merlin 623 viz: Two-speed, two-stage supercharger and increased boost for take-off and climb In addition, full depth intercooling for take-off and maximum continuous power conditions with half depth intercooling and automatic control charge heating for normal cruise is introduced Standard English splines on propeller shaft
Merlin 626	Basic type	12.0/ 10.1	5.79/ 7.06	.42	+20½	–	No production Similar to Merlin 624 viz: Two-speed, two-stage supercharger and increased boost for take-off and climb In addition, full depth intercooling for take-off and maximum continuous power conditions with half depth intercooling, and automatic control charge heating for normal cruise is introduced Special installation features including American type splines on the propeller for D.C.4 aircraft, as Merlin 620 and 624 type engines

ROLLS-ROYCE PISTON ENGINES
LIST OF COMMERCIAL MARK NUMBERS

Mark Number	Customer	Supercharger		Propeller Reduction Gear Ratio	Max. Boost lb/sq.inch. 100/130 Grade Fuel	Aircraft Type	Remarks
		Rotor Dia.	Gear Ratios				
Merlin 626-1	B.O.A.C.	12.0/10.1	5.79/7.06	.42	+20½	D.C.4	As Merlin 626 type engines Now 724-1 and 724-1C
Merlin 626-12	C.P.A.	12.0/10.1	5.79/7.06	.42	+20½	D.C.4	As Merlin 626 type engines
Merlin 630		12.0/10.1	5.79/7.06	.42	+20		Similar to Merlin 620 but with 7:1 compression ratio. No production
Merlin 631		12.0/10.1	5.79/7.06	.42	+20		Similar to Merlin 621 with 7:1 compression ratio. No production
Merlin 640		12.0/10.1	5.79/7.06	.471	+19		Similar to Merlin 620 but with 6.3:1 compression ratio. No production
Merlin 641		12.0/10.1	5.79/7.06	.42	+19		Similar to Merlin 621 but with 6.3:1 compression ratio. No production
Merlin 724-1	B.O.A.C.	12.0/10.1	5.79/7.06	.42	+20½	D.C.4	Similar to Merlin 626-1 but incorporating Merlin Mod.2989- Intercooling system segregated from main coolant system to give full or no intercooling

ISSUED BY PRODUCTION DRAWING OFFICE • ROLLS-ROYCE LIMITED DERBY   DATE ISSUED   FEBRUARY 1953

Mark Number	Customer	Supercharger		Propeller Reduction Gear Ratio	Max. Boost lb/sq/inch. 100/130 Grade Fuel	Aircraft Type	Remarks
		Rotor Dia.	Gear Ratios				
Merlin 724-1C	B.O.A.C.	12.0/ 10.1	5.79/ 7.06	.42	+20½	D.C.4	As Merlin 724-1 but incorporating Merlin Mod.4300- Crossover exhaust system
Merlin 722-10	Canadair	12.0/ 10.1	5.79/ 7.06	.42 .457	+20½		As Merlin 622-10 but incorporating Merlin Mod.2676- Makes provision for full intercooling by connecting the intercooler elements in series
Merlin 724-10	Trans Canada Airways	12.0/ 10.1	5.79/ 7.06	.4707 .42	+20½	D.C.4	As Merlin 624-10 but incorporating Merlin Mod.2676- Makes provision for full intercooling by connecting the intercooler elements in series

ISSUED BY PRODUCTION DRAWING OFFICE · ROLLS-ROYCE LIMITED DERBY

DATE ISSUED FEBRUARY 1953

Page CM.8

M E R L I N.

Two Speed – Two Stage Supercharger Series.
BASIC ENGINE FOR ALL THE FOLLOWING MARKS.
For Performance Data see Bulletin 15.

R'dn. Gear Ratio.	Supercharger. Rotor Dia.	Gear Ratio.	Type of Carb. or Fuel Injection Pump.	Coolant Direction of Flow.	Starter.	Remarks.
0·42	11·5″ 10·1″	5·52 8·41	S.U. Float Feed and R.A.E. Anti-G	Normal	Electric	**R'dn. Gear.** Universal Prop. Shaft. **Cyl. Block.** Two-piece. Pressure water/glycol cooled, Pressure cabin blower fitted. **Intercooler.** Fitted.

Mark No.	Aircraft.	Variations from Basic.	Mark No.	Aircraft.	Variations from Basic.
60	Wellington VI	**Cyl. Block.** One-piece. Limited production.	73	Mosquito IX, XIV, XVI	**S C. Gear Ratio.** 6·39 8·03 **Coolant Direction of Flow.** Reversed.
61	Spitfire IX F	**S C. Gear Ratio.** 6·39 8·03	76	Mosquito IX, XV, XVI	**S C. Rotor Dia.** 12·0″ 10·1″ **S C. Gear Ratio.** 6·39 8·03 **Carb.** Bendix Stromberg injection type 8D.44/1. **Coolant Direction of Flow.** Reversed. **Cabin Blower.** Not fitted.
62	Wellington	Limited production.			
63	Spitfire VIII, IX, XI	**R'dn. Gear Ratio.** 0·477 **S/C. Gear Ratio.** 6·39 8·03 **S/C.** Strengthened drive shaft. **Cabin Blower.** Not fitted.	77	Mosquito IX, XV, XVI	**S C. Rotor Dia.** 12·0″ 10·1″ **S C. Gear Ratio.** 6·39 8·03 **Coolant Direction of Flow.** Reversed. **Carb.** Bendix Stromberg injection type 8D.44/1.
63A	Spitfire VIII, IX	**C'Case.** Merlin 64 type but with cabin blower drive parts deleted and blanking covers fitted. **R'dn. Gear Ratio.** 0·477 **S/C. Gear Ratio.** 6·39 8·03 **Cabin Blower.** Not fitted.	85	Lincoln I	**S C. Rotor Dia.** 12·0″ 10·1″ **S C. Gear Ratio.** 5·79 7·06 **Carb.** Bendix Stromberg injection type 8D.44 3. Introduction of Aux. gearbox drive (30 hp). Intercooler header tank integral with intercooler. **Cabin Blower.** Not fitted.
64	Spitfire VII, X	**R'dn. Gear Ratio.** 0·477 **S/C. Gear Ratio.** 6·39 8·03 Similar to Merlin 63 but with drive for cabin blower.			
66	Spitfire VIII, IX, LF	**R'dn. Gear Ratio.** 0·477 **S/C. Rotor Dia.** 12·0″ 10·1″ **S/C. Gear Ratio.** 5·79 7·06 **Carb.** Bendix Stromberg injection type 8D.44/1. **Controls.** Interconnected. **Cabin Blower.** Not fitted. NOTE. Certain Merlin 66 engines, special order only, have been converted to operate at +25 lb. boost combat rating (Mod. 785).	85A	Lincoln I	**S C. Rotor Dia.** 12·0″ 10·1″ **S C. Gear Ratio.** 5·79 7·06 **Carb.** Bendix Stromberg injection type 8D.44 3. Auxiliary gearbox drive (30 hp). **Coolant Pump.** Merlin 100 type (ball bearing). **Joints.** Deletion of main structural joints. **Cabin Blower.** Not fitted.
66–8	Spitfire (Turkish Government)	**R'dn. Gear Ratio.** 0·477 **S/C. Rotor Dia.** 12·0″ 10·1″ **S/C. Gear Ratio.** 5·79 7·06 **Carburation.** Bendix Stromberg injection type 8D.44/1.	85B	Lincoln I	**S C. Rotor Dia.** 12·0″ 10·1″ **S C. Gear Ratio.** 5·79 7·06 **Carb.** Bendix Stromberg injection type 8D.44 3. Aux. gearbox drive (30 hp). **Dual Drive.** Modified for fitting of American 4G8 C.S.U. (built for Australian Contract). **Cabin Blower.** Not fitted.
70	Spitfire VIII, IX, XI, HF.	**R'dn. Gear Ratio.** 0·477 **S/C. Rotor Dia.** 12·0″ 10·1″ **S/C. Gear Ratio.** 6·39 8·03 **Carb.** Bendix Stromberg injection type 8D.44/1. **Cabin Blower.** Not fitted.	86	Lincoln	**S C. Rotor Dia.** 12·0″ 10·1″ **S C. Gear Ratio.** 5·79 8·03 **Carb.** Bendix Stromberg injection type 8D.44/3. Aux. gearbox drive (30 hp). Built for high altitude with S/C anti-surge diffuser ring. **Cabin Blower.** Not fitted.
72	Mosquito VIII, IX, XV, XVI	**S/C. Gear Ratio.** 6·39 8·03 **Coolant Direction of Flow.** Reversed. **Cabin Blower.** Not fitted.			

MERLIN.
Two-Speed Two-Stage Merlin 100 Series.
BASIC ENGINE FOR ALL THE FOLLOWING MARKS.
For Performance Data see Bulletin 15.

R'dn. Gear Ratio.	Supercharger.		Type of Carb. or Fuel Injection Pump.	Coolant Direction of Flow.	Starter.	Remarks.
	Rotor Dia.	Gear Ratio.				
.042	12.0″ 10.1″	5.79 7.06	S.U. Fuel injection pump with de-aerator.	Normal	Electric	**C'Case.** Strengthened, universal for end to end oil feed. **R'dn. Gear.** Strengthened casing and pinion bearings. **S/C.** Overhung first stage rotor with centrifugal fuel spray. **S/C. Intake.** Improved short elbow with mechanical accelerator pump and fuel injection features. **Cyl. Block.** Strengthened with rocker cover breathers. **Coolant Pumps.** Double packless gland oil lubricated ball bearings. **Starter Motor Drive.** Simplified single layshaft. **Aux. Gearbox Drive.** Fitted or provision.

Mark No.	Aircraft.	Variations from Basic.	Mark No.	Aircraft.	Variations from Basic.
102	Tudor (Prototype)	**Aux. Gearbox Drive.** Fitted (60-80 hp) Limited Production.	131	Hornet F.1 (Stbd. F.3 Engine)	Merlin 131 is same as Merlin 130 with **Prop. Shaft.** Opposite direction of rotation.
102A	Tudor (Prototype)	Incorporating an after-heater. **Aux. Gearbox Drive.** Fitted (60-80 hp) Limited Production.	132	Sea Hornet (Port Engine)	As for Merlin 130 except for General purposes with internal oil feed for braking propeller. Downdraught with Corliss throttle. Very limited production.
104	Mosquito 32	**S/C.** Merlin 76 type, first stage ditfuser ring. **A.B.C.** Cleared for 25 lb/sq. in. boost. **Coolant Direction of Flow.** Reversed.			
113	Mosquito 32, 34, 35, 36	**S/C. Gear Ratio.** 6.39 8.03 **Coolant Direction of Flow.** Reversed.	133	Sea Hornet (Stbd. Engine)	Same as Merlin 132 except for :— **R'dn. Gear Shaft.** Left hand rotation. Very limited production.
113A	Mosquito 32, 34, 35, 36	**S/C. Gear Ratio.** 6.39 8.03 **S/C.** Anti-surge diffusers fitted— Merlin 76 type. **Coolant Direction of Flow.** Reversed.	134	Sea Hornet PR. 22 F. XX NF. 21	Same as Merlin 130 but **Controls.** Corliss throttle and fixed ignition.
114	Mosquito 32, 34, 35, 36	**S/C. Gear Ratio.** 6.39 8.03 **Coolant Direction of Flow.** Reversed. **Pressure Cabin Blower.** Fitted.	134-5	Nancu Type Argentine Government	Same as Merlin 134.
114A	Mosquito 32, 34, 35, 36	**S/C. Gear Ratio.** 6.39 8.03 **S/C.** Anti-surge diffuser fitted, Merlin 76 type. **Coolant Direction of Flow.** Reversed. **Pressure Cabin Blower.** Fitted.	135	Sea Hornet PR. 22 F. XX NF. 21	Same as Merlin 134 but with **Reduction Gear Shaft.** Left hand.
130	Hornet F.1 (Port F.3 Engine)	**Crankcase.** Strengthened, with internal oil feed. **S/C. Change Mechanism.** Built in pneumatic ram. **S/C. Intake.** Downdraught with Vortex or Corliss throttles. **A.B.C.** Mounted in reverse position on engine. Cleared for 25 lb/sq. in. boost. **Coolant Pump.** Griffon 65 type mounted on intercooler pump bracket. **Coolant Direction of Flow.** Reversed. **Controls.** To suit De Havilland Hornet.	135-5	Nancu Type. Argentine Government	Same as Merlin 135.
			140	Sturgeon II	**R'dn. Gear.** Contra rotating prop. shafts, both rotating 0.512 of C/S. **Coolant.** Shunt System. **Aux. Gearbox Drive.** Fitted (60-80 hp) **Starter.** Coffman. Cleared for 25 lb/sq. in. boost.

MERLIN CIVIL AND COMMERCIAL TYPES.

Two Speed – Two Stage Supercharger Series.

BASIC ENGINE FOR THE FOLLOWING MARKS.

For Performance Data see Bulletin 15.

R'dn. Gear Ratio.	Supercharger. Rotor Dia.	Gear Ratio.	Type of Carb. or Fuel Injection Pump.	Coolant Direction of Flow.	Starter.	Remarks.
0·420	12·0" 10·1"	5·79 7·06	S.U. Injection pump with de-aerator	Normal	Electric	**Cyl. Block.** Two-piece. Pressure water glycol cooled. **Induction System.** Combined intercooler and after-heater. **Controls.** Corliss type throttle. **Aux. Gearbox.** 100 h.p. type. **Rocker Cover Breathing.** Built-in.

Mark No.	Aircraft and Operator.	Variations from Basic.	Mark No.	Aircraft and Operator.	Variations from Basic.
600	Prototype Tudor	Converted 102A. **Induction System.** Incorporating an after-heater. **Aux. Gearbox Drive.** Fitted 60–80 hp). **Controls.** M.100 series throttle.	622	**D.C.4. M.2.** T.C.A. North Star Domestic (Internal Airlines)	As Merlin 620 but with increased boost for take-off and climb. **R'dn. Gear Ratio.** 0·4707
620	**D.C.4. M.1.** T.C.A. and Royal Canadian Air Force	**R'dn. Gear Ratio.** 0·4707 American type splines on prop. shaft. Special installation features for D.C.4 aircraft.	623–2	**Tudor IV**	Similar to Merlin 621 type engines, but with increased boost for take-off and climb.
621–1	**Tudor II**	Special installation features to suit this type of Tudor aircraft.	624	**D.C. 4 M.2.** T.C.A. North Star (Atlantic)	As Merlin 622 except reduction gear ratio 0·420
621–2	**Tudor II, IV and V**	Special installation features to suit these types of aircraft.	626–1	**Canadair IV C.4.** Argonaut Class (B.O.A.C.)	Similar to Merlin 624 type engines, *viz.*, increased boost for take-off and climb, full depth intercooling for take-off and maximum continuous power conditions with half depth intercooling, automatic control charge heating for normal cruise. Special installation features including American type splines on the propeller for D.C.4 A.C. as Merlin 620 and 624 type engines. Suitable for application of R.R. Power meters.
621–15	**Lincoln** (Argentine Government)	As Merlin 621 but with installation features for Lincoln aircraft.	626–12	**Canadair IV C.4** Canadian Pacific Airways, Empress Class	Similar to Merlin 626–1.

Mark No.	Aircraft and Operator	Variations from Basic	Mark No.	Aircraft and Operator	Variations from Basic
722-10	D.C.4 M.2 T.C.A. North Star Domestic (Internal Airlines)	As Merlin 622-10 but incorporating Merlin Mod.2676—Makes provision for full intercooling by connecting the intercooler elements in series.	724-1C	D.C.4 M.2 Argonaut Class. (B.O.A.C.)	As Merlin 724-1 but incorporating Merlin Mod.4300—Crossover Exhaust System.
724-1	D.C.4. M.2 Argonaut Class. (B.O.A.C.)	Similar to Merlin 626-1 but incorporating Merlin Mod.2989—Intercooling system segregated from main coolant system to give full or no intercooling.	724-10	D.C.4. M.2 T.C.A. North Star (Atlantic)	As Merlin 624-10 but incorporating Merlin Mod. 2676—Makes provision for full intercooling by connecting the intercooler elements in series.

190

A M E R I C A N - B U I L T   M E R L I N.

Two Speed – Two Stage Supercharger Series.

BASIC ENGINE FOR ALL THE FOLLOWING MARKS.

R'dn. Gear Ratio.	Supercharger. Rotor Dia.	Supercharger. Gear Ratio.	Type of Carb. or Fuel Injection Pump.	Coolant Direction of Flow.	Starter.	Remarks.
0·479	12·0″ 10·1″	6·39 8·095	Bendix Stromberg injection type. PD. 16/B1	Normal	Electric	**R'dn. Gear. Prop. Shaft.** Universal. **Cyl. Block.** Two-piece. Pressure water/glycol cooled.

Mark No.	Aircraft.	Variations from Basic.
V.1650 -3	Mustang III	**R'dn. Gear. Prop. Shaft.** S.A.E. American splines. **Coolant Pump.** Ball bearing type with packless gland. **Intercooler.** Mounted on cylinder blocks. **Oil Filter.** Cuno oil pressure filter between oil pump and oil relief valve box.
V.1650 -7	Mustang III, IV	**R'dn. Gear. Prop. Shaft.** American splines. **S/C. Gear Ratios.** 5·80 7·34 **NOTE.** Certain V.1650–7 engines have been modified to operate at +25 lb. boost combat rating. Special order only.
V.1650 -9	Mustang	**R'dn. Gear. Prop. Shaft.** American splines. **Carburation.** Water methanol injection unit. **Boost Control.** Simmonds power control unit.
V.1650 -9A	Mustang	**R'dn. Gear. Prop. Shaft.** American splines. **Boost Control.** Simmonds power control unit.
V.1650 -11	Projected for Mustang	**R'dn. Gear. Prop. Shaft.** American splines. **Carburation.** Speed density pump. Water Methanol injection unit. **Boost Control.** Simmonds power control unit. (Experimental only).

Mark No.	Aircraft.	Variations from Basic.
V.1650 -13	Projected for Mustang	**R'dn. Gear. Prop. Shaft.** American splines. **Boost Control.** Simmonds power control unit. (Experimental only)
V.1650 -17	Projected for Mustang	**R'dn. Gear. Prop. Shaft.** American splines. **S/C. Gear Ratios.** 5·80 7·34 **Boost Control.** Simmonds power control unit. (Experimental only)
V.1650 -21	N.A.A. Experimental Fighter XP. 82	**R'dn. Gear. Prop. Shaft.** Reverse rotation. American splines. **Carb.** Speed density pump.
68	Lincoln B.1. B.2. B.2 (3A) B.2 (49)	**R'dn. Gear Ratio.** 0·42 **S/C. Gear Ratios.** 5·80 7·34 **Aux. Gearbox Drive.** Fitted. American version of Merlin 85. **Carb.** Type PD. 18.B1.
68A	Lincoln B.1. B.2. B.2 (3A) B.2 (4G)	Modified Merlin 68. **R'dn. Gear Ratio.** 0·42 **S/C. Gear Ratios.** 5·80 7·34 **Intercooler.** Incorporates after-heater with automatic charge control. **Coolant Pump.** Merlin 100 type (ball bearing). **Carb.** Throttle stops to prevent surge. Type PD. 18D1A or B1. **Aux. Gearbox Drive.** Fitted. **Joints.** Deletion of main structural joints. Fixed ignition.
69	Mosquito	**R'dn. Gear Ratio.** 0·42 **S/C. Gear Ratios.** 5·80 7·34 **Coolant Direction of Flow.** Reversed. American Version of Merlin 67. **Carb.** Type PD. 18 B1.
266	Spitfire IX, XVI	**S/C. Gear Ratios.** 5·80 7·34 **Intercooler.** Integral header tank type. American-built Merlin 66. **Carb.** Type PD. 18 B1.

191

PACKARD MERLIN 100 SERIES

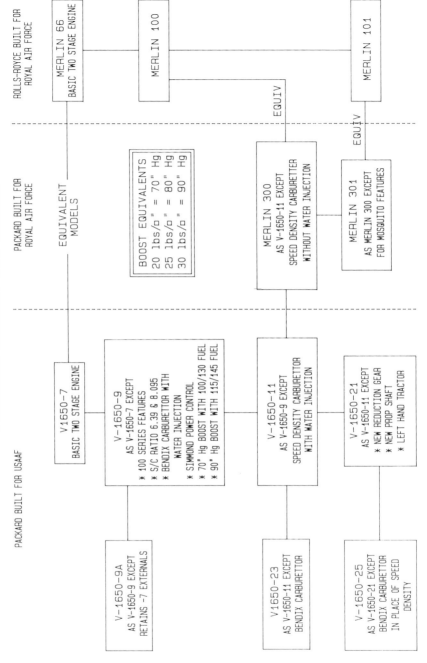

ROLLS-ROYCE BUILT FOR
ROYAL AIR FORCE

MERLIN 66
BASIC TWO STAGE ENGINE

MERLIN 100

MERLIN 101

PACKARD BUILT FOR
ROYAL AIR FORCE

EQUIVALENT
MODELS

EQUIV

EQUIV

BOOST EQUIVALENTS
20 lbs/□" = 70" Hg
25 lbs/□" = 80" Hg
30 lbs/□" = 90" Hg

MERLIN 300
AS V-1650-11 EXCEPT
SPEED DENSITY CARBURETTER
WITHOUT WATER INJECTION

MERLIN 301
AS MERLIN 300 EXCEPT
FOR MOSQUITO FEATURES

PACKARD BUILT FOR USAAF

V1650-7
BASIC TWO STAGE ENGINE

V-1650-9
AS V-1650-7 EXCEPT
* 100 SERIES FEATURES
* S/C RATIO 6.39 & 8.095
* BENDIX CARBURETTOR WITH
  WATER INJECTION
* SIMMOND POWER CONTROL
* 70" Hg BOOST WITH 100/130 FUEL
* 90" Hg BOOST WITH 115/145 FUEL

V-1650-11
AS V-1650-9 EXCEPT
SPEED DENSITY CARBURETTOR
WITH WATER INJECTION

V-1650-21
AS V-1650-11 EXCEPT
* NEW REDUCTION GEAR
* NEW PROP SHAFT
* LEFT HAND TRACTOR

V-1650-9A
AS V-1650-9 EXCEPT
RETAINS -7 EXTERNALS

V1650-23
AS V-1650-11 EXCEPT
BENDIX CARBURETTOR

V-1650-25
AS V-1650-21 EXCEPT
BENDIX CARBURETTOR
IN PLACE OF SPEED
DENSITY

MISCELLANEOUS ILLUSTRATIONS

LEADING PARTICULARS

MERLIN Mk. 113, 113A, 114, 114A, 130, 131, 134 and 135 AERO-ENGINES

GENERAL

Type of engine	Supercharged, geared, pressure-liquid cooled V-engine, fitted with two-speed, two-stage, liquid-cooled supercharger with intercooler
Number of cylinders	12

Arrangement of cylinders
Two banks of six cylinders with an included angle of 60 deg.

Cylinder numbering	Propeller	1A, 2A, 3A, 4A, 5A, 6A
		1B, 2B, 3B, 4B, 5B, 6B

Bore	5·4 in.
Stroke	6·0 in.
Swept volume...	1,648 cu. in.
Compression ratio	6·0 to 1

Direction of rotation

Crankshaft	Left-hand

Propeller shaft

Merlin Mk. 113, 113A, 114, 114A, 130, 134	Right-hand
Merlin Mk. 131 and 135	Left-hand

Supercharger

Type	Two-speed, two-stage

Gear ratios

Merlin Mk. 130, 131, 134 and 135	5·79 to 1 and 7·06 to 1
Merlin Mk. 113, 113A, 114 and 114A	6·39 to 1 and 8·03 to 1

Propeller reduction gear

Type	Single spur reduction

Gear ratio

Merlin, Mk. 113, 113A, 114, 114A, 130 and 134	0·420 to 1
Merlin, Mk. 131 and 135	0·422 to 1

Weight of engine, nett dry (approximate)

Merlin, Mk. 113, 113A, 114, 114A	1,670 lb.
Merlin Mk. 130, 131, 134	1,730 lb.
Merlin, Mk. 135	1,775 lb.

PERFORMANCE

International power rating—

	Low gear	High gear
Merlin Mk. 130, 131, 134 and 135	1,410 b.h.p. at 10,000 ft.	1,315 b.h.p. at 20,500 ft.
Merlin Mk. 113, 113A, 114 and 114A...	1,380 b.h.p. at 15,750 ft.	1,200 b.h.p. at 29,750 ft.

Combat power rating—

Merlin Mk. 130, 131, 134 and 135	1,830 b.h.p. at 5,500 ft.	1,690 b.h.p. at 18,000 ft.
Merlin Mk. 113, 113A, 114 and 114A...	1,690 b.h.p. at 13,000 ft.	1,435 b.h.p. at 27,250 ft.

OIL

Type	Oil OM–270
	(A.P.1464C, Vol. 2, Part 1, leaflet No. 4)
Consumption at maximum cruising conditions	6 to 20 pints per hour

Pressure (main)

Minimum in flight	See Operating Limitations

IGNITION

Firing order *1A, 6B, 4A, 3B, 2A, 5B, 6A, 1B, 3A, 4B, 5A, 2B*

Magnetos—
Number ... ...   *Two*
Type ... ...   *B.T.H. C6S.E. 125/1A, or /2, or*
Rotax N.S.E. 12/7, or N12A, or N12/B

Direction of rotation (looking on drive end) ... ...   *Port, clockwise*
Starboard, anti-clockwise

Speed of rotation ... ...   *1·5 crankshaft speed*

Contact breaker gap ... ...   *0·012 in. ± 0·001 in.*

Timing
Port (exhaust plugs) ... ...   *45 deg. before T.D.C.*
Starboard (inlet plugs) ... ...   *38 deg. before T.D.C.*
Sparking plug types ... ...   *R.C.5/4, R.C.5/7, L.R.3.R, K.R.3.R.*

Sparking plug gaps	Max. operating altitude
·018 in. to ·021 in.	30,000 ft.
·015 in. to ·018 in.	40,000 ft.
·012 in. to ·015 in.	45,000 ft.
·012 in.	45,000 ft.
	and above

CARBURATION

Injection pump ... ...   *S.U. single point*
Type *6,000/601*
Fuel ... ...   *100/130 AVGAS (Stores Ref. 34A/75)*
Maximum fuel demand ... ...   *205 gallons per hour*
Fuel pressure warning light setting ... ...   *3 lb. per sq. in.*

VALVES

Valve timing (with 0.014 in. cam clearance)

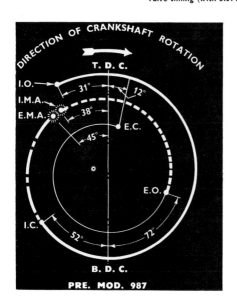
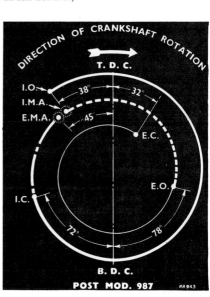

Timing diagram, post Mod. 2230

Valve clearances (measured between cam and rocker)

Inlet	*0·009 in.*
Exhaust	*0·015 in.*
Tolerance after running	*−0·003 in.*

COOLANT SYSTEM

Circulation

Main system ...	"*Reverse*" *flow (Header tank—pump—radiator—cylinders)*
Intercooler system (Merlin Mk. 113 to 114A)	*Header tank — pump — radiator — supercharger — intercooler*
(Merlin Mk. 130 to 134)	*Header tank — pump — supercharger — radiator — intercooler*
Mixture for normal operation ...	*70% water, 30% inhibited ethylene glycol (Spec. D T.D. 344A)*
Mixture for cold weather operation	*55% water, 45% inhibited ethylene glycol (Spec. D.T.D. 344A)*

STARTING SYSTEM

Type	*Electric turning gear, B.T.H. type CA.4750 model FE.3, FE.4, or FE.4/1*

PROPELLER

Type	*De Havilland hydromatic fully-feathering with double-acting constant-speed control*

ACCESSORIES

The following accessories and ancillary equipment are normally fitted. Accessories used in any particular installation are specified in the appropriate power plant or aircraft Air Publication.

Accessories	Speed ratio relative to crankshaft	Direction of rotation looking on driving spindle of accessory
Constant speed governor unit	0·828	Clockwise
Vacuum pump	0·828	Clockwise
*Electric generator	1·953	Anti-clockwise
†Electric generator	2·1	Anti-clockwise
‡Electric generator	2·11	Clockwise
Starter motor	86·5	Clockwise
Air compressor (camshaft)	0·500	Clockwise
Hydraulic pump (camshaft)	1·0	Clockwise
Engine speed indicator drive (camshaft) ...	0·25	Clockwise
Hydraulic pump (crankcase)	0·992 or 0·502	Clockwise
§Cabin supercharger	0·913	Anti-clockwise

* Merlin, Mk. 113, 113A, 114 and 114A   ‡ Merlin Mk. 131 and 135
† Merlin Mk. 130 and 134   § Merlin Mk. 114 and 114A

THE 'RM17SM' ENGINE

Production Mark Number   Not allocated

M.O.S. Experimental
Nomenclature   RM17SM

Supercharger
    Rotor diameters   12.7 in. and 10.7 in.
    Gear Ratios   5.79 and 7.06

Reduction Gear Ratio   0.42

Maximum Boost   30 lb/sq.in.

Fuel   RDE/F/290 (115/150 grade)

Ignition timing   38/45° B.T.D.C.

Camshaft   D 24196 (increased exhaust duration)

Nominal Valve timing   Inlet open 36° B.T.D.C.
          Inlet close 68° A.B.D.C.
          Exhaust open 76° B.B.D.C.
          Exhaust close 48° A.T.D.C.

Sparking Plugs   RC 5/3 and RC 5/5

Injection Equipment   Rolls Royce Injection pump-single point

Power   2200 hp at 2000 ft -MS/2100hp at 15000 ft - FS

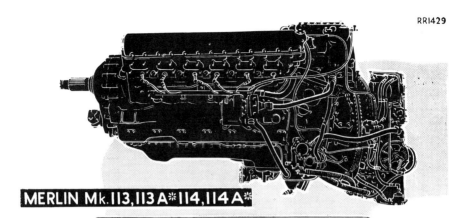

MERLIN Mk. 113, 113 A✷114, 114 A✷

ENGINE DATA

SUPERCHARGER: GEAR RATIOS 6·39 AND 8·03 TO 1

CARBURATION: UPDRAUGHT INTAKE

PROPELLER REDUCTION GEAR: GEAR RATIO 0·420 TO 1

PROPELLER SHAFT ROTATION: RIGHT HAND

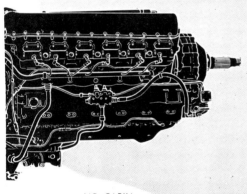

NO CABIN
SUPERCHARGER DRIVE

CABIN
SUPERCHARGER
DRIVE

MERLIN Mk. 113, 113 A✷

MERLIN Mk. 114, 114 A✷

✷ THE SUFFIX "A" INDICATES A LATER MOD STANDARD

ENGINE IDENTIFICATION CHART

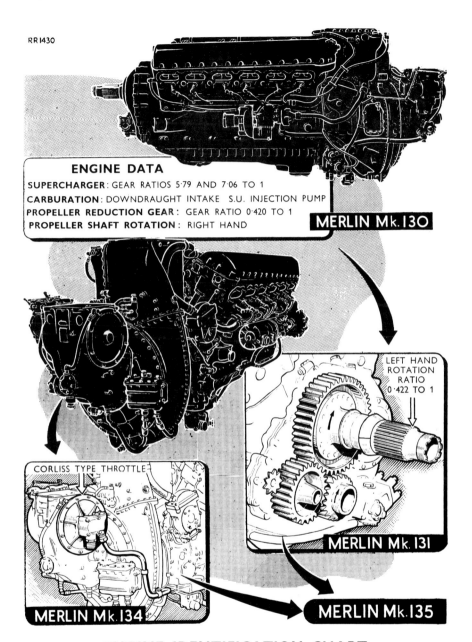

ENGINE DATA

SUPERCHARGER: GEAR RATIOS 5·79 AND 7·06 TO 1
CARBURATION: DOWNDRAUGHT INTAKE S.U. INJECTION PUMP
PROPELLER REDUCTION GEAR: GEAR RATIO 0·420 TO 1
PROPELLER SHAFT ROTATION: RIGHT HAND

MERLIN Mk. 130

LEFT HAND
ROTATION
RATIO
0·422 TO 1

CORLISS TYPE THROTTLE

MERLIN Mk. 131

MERLIN Mk. 134

MERLIN Mk. 135

ENGINE IDENTIFICATION CHART

GEAR TRAIN DIAGRAM

LOW GEAR FRICTION CLUTCH

STARTER MOTOR 86·5

SUPERCHARGER GEAR RATIOS

| MERLIN MK.130,131,134 & 135 | 5·79 ... 7·06 |
| MERLIN MK.113,113A,114,114A | 6·39 ... 8·03 |

DRIVE FOR ACCESSORY 0·5

HIGH GEAR FRICTION CLUTCHES

DRIVE FOR ACCESSORY 1·0

DRIVE FOR ACCESSORY 0·792

FUEL PUMP DRIVE 0·917

MERLIN COOLANT PUMP 1·5 MK.113, 113A, 114 AND 114 A

R.P.M. INDICATOR DRIVE 0·25

MAGNETO 1·5

PRESSURE PUMP 0·738

REAR SCAVENGE PUMP 0·738

CAMSHAFTS 0·5

GENERATOR :—
MERLIN MK.130 AND 134, 2·1
MERLIN MK.131 AND 135 : 2·11
OR
CABIN SUPERCHARGER DRIVE:—
MERLIN MK. 114 AND 114 A ONLY : 0·913

FRONT SCAVENGE PUMP 0·738

COOLANT PUMP MERLIN MK.130,131 134,135 : 1·5
OR
GENERATOR 1·953
MERLIN MK.113, 113A,114,114 A

UNDERCARRIAGE HYDRAULIC PUMP MERLIN MK. 113, 113A, 114 AND 114 A ONLY — 0·502 OR 0·992

INTERCOOLER PUMP 1·497

CRANKSHAFT 1·0

REDUCTION GEAR RATIOS

| MERLIN MK. 113,113A,114, 114A AND 134 | 0·420 |
| MERLIN MK.131 AND 135 | 0·422 |

VACUUM PUMP 0·828

PROPELLER CONSTANT SPEED UNIT 0·828

200

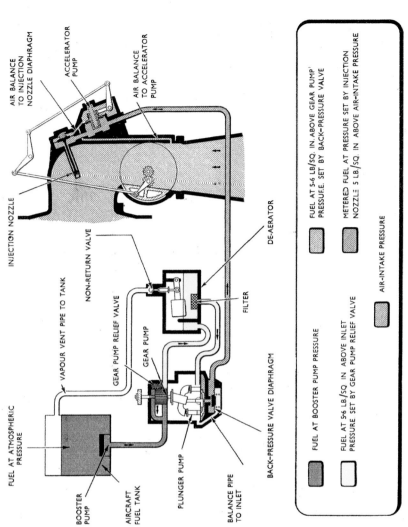

Merlin 140 – Engine fuel system diagram showing the Corliss Throttle

AIR BALANCE TO INJECTION NOZZLE DIAPHRAGM

ACCELERATOR PUMP

AIR BALANCE TO ACCELERATOR PUMP

INJECTION NOZZLE

NON-RETURN VALVE

DE-AERATOR

VAPOUR VENT PIPE TO TANK

GEAR PUMP RELIEF VALVE

GEAR PUMP

FILTER

FUEL AT ATMOSPHERIC PRESSURE

BOOSTER PUMP

AIRCRAFT FUEL TANK

PLUNGER PUMP

BALANCE PIPE TO INLET

BACK-PRESSURE VALVE DIAPHRAGM

FUEL AT BOOSTER PUMP PRESSURE

FUEL AT 5·6 LB/SQ IN ABOVE INLET PRESSURE SET BY GEAR PUMP RELIEF VALVE

FUEL AT 5·6 LB/SQ. IN. ABOVE GEAR PUMP PRESSURE. SET BY BACK-PRESSURE VALVE

METERED FUEL AT PRESSURE SET BY INJECTION NOZZLE 5 LB/SQ. IN. ABOVE AIR-INTAKE PRESSURE

AIR-INTAKE PRESSURE

201

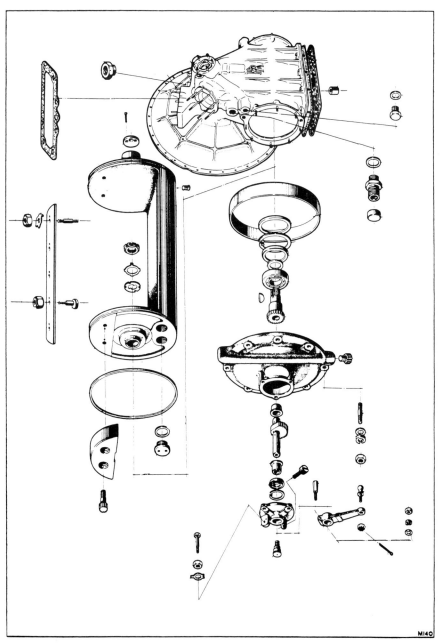

M140

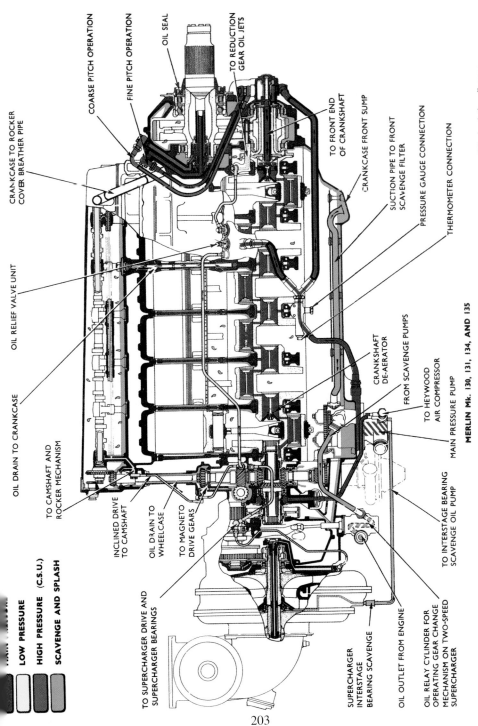

COARSE PITCH OPERATION

FINE PITCH OPERATION

OIL SEAL

TO REDUCTION GEAR OIL JETS

CRANKCASE TO ROCKER COVER BREATHER PIPE

TO FRONT END OF CRANKSHAFT

CRANKCASE FRONT SUMP

SUCTION PIPE TO FRONT SCAVENGE FILTER

PRESSURE GAUGE CONNECTION

THERMOMETER CONNECTION

OIL RELIEF VALVE UNIT

OIL DRAIN TO CRANKCASE

TO CAMSHAFT AND ROCKER MECHANISM

CRANKSHAFT DE-AERATOR

FROM SCAVENGE PUMPS

TO HEYWOOD AIR COMPRESSOR

MAIN PRESSURE PUMP

INCLINED DRIVE TO CAMSHAFT

OIL DRAIN TO WHEELCASE

TO MAGNETO DRIVE GEARS

TO INTERSTAGE BEARING SCAVENGE OIL PUMP

TO SUPERCHARGER DRIVE AND SUPERCHARGER BEARINGS

LOW PRESSURE

HIGH PRESSURE (C.S.U.)

SCAVENGE AND SPLASH

SUPERCHARGER INTERSTAGE BEARING SCAVENGE

OIL OUTLET FROM ENGINE

OIL RELAY CYLINDER FOR OPERATING GEAR CHANGE MECHANISM ON TWO-SPEED SUPERCHARGER

Oil circulation diagram

MERLIN Mk. 130, 131, 134, AND 135

203

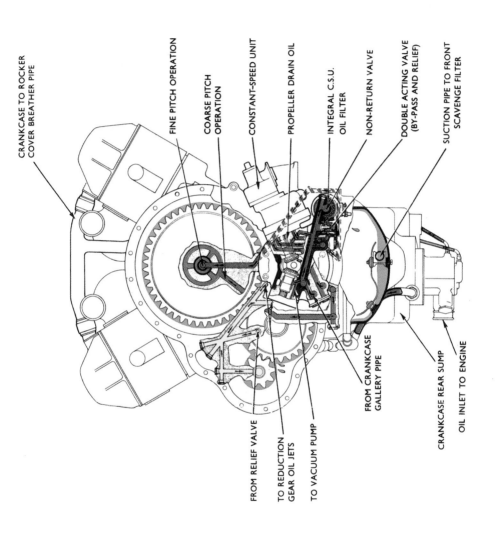

CRANKCASE TO ROCKER COVER BREATHER PIPE

FINE PITCH OPERATION

COARSE PITCH OPERATION

CONSTANT-SPEED UNIT

PROPELLER DRAIN OIL

INTEGRAL C.S.U. OIL FILTER

NON-RETURN VALVE

DOUBLE ACTING VALVE (BY-PASS AND RELIEF)

SUCTION PIPE TO FRONT SCAVENGE FILTER

FROM RELIEF VALVE

TO REDUCTION GEAR OIL JETS

TO VACUUM PUMP

FROM CRANKCASE GALLERY PIPE

CRANKCASE REAR SUMP

OIL INLET TO ENGINE

FRONT SECTIONAL VIEW (MERLIN Mk. 135)

204

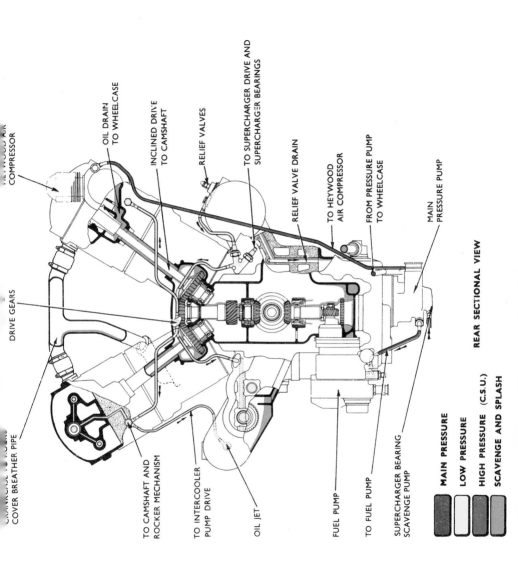

COVER BREATHER PIPE

DRIVE GEARS

COMPRESSOR

OIL DRAIN
TO WHEELCASE

INCLINED DRIVE
TO CAMSHAFT

RELIEF VALVES

TO SUPERCHARGER DRIVE AND
SUPERCHARGER BEARINGS

RELIEF VALVE DRAIN

TO HEYWOOD
AIR COMPRESSOR

FROM PRESSURE PUMP
TO WHEELCASE

MAIN
PRESSURE PUMP

TO CAMSHAFT AND
ROCKER MECHANISM

TO INTERCOOLER
PUMP DRIVE

OIL JET

FUEL PUMP

TO FUEL PUMP

SUPERCHARGER BEARING
SCAVENGE PUMP

REAR SECTIONAL VIEW

MAIN PRESSURE

LOW PRESSURE

HIGH PRESSURE (C.S.U.)

SCAVENGE AND SPLASH

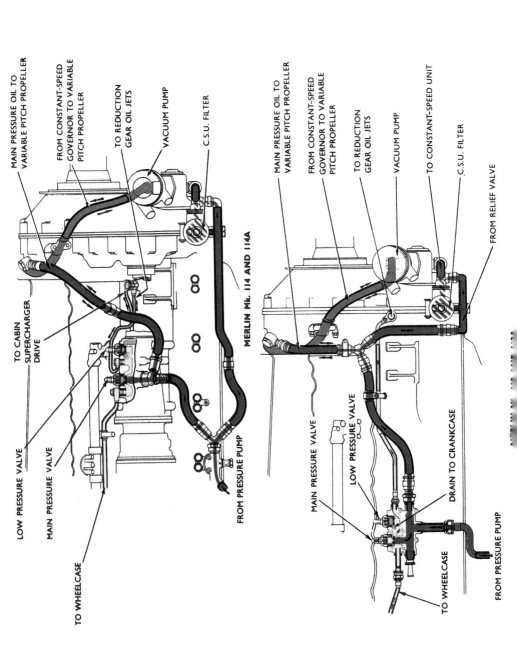

MAIN PRESSURE OIL TO VARIABLE PITCH PROPELLER

FROM CONSTANT-SPEED GOVERNOR TO VARIABLE PITCH PROPELLER

TO REDUCTION GEAR OIL JETS

VACUUM PUMP

C.S.U. FILTER

TO CABIN SUPERCHARGER DRIVE

LOW PRESSURE VALVE

MAIN PRESSURE VALVE

TO WHEELCASE

FROM PRESSURE PUMP

MERLIN Mk. 114 AND 114A

MAIN PRESSURE OIL TO VARIABLE PITCH PROPELLER

FROM CONSTANT-SPEED GOVERNOR TO VARIABLE PITCH PROPELLER

TO REDUCTION GEAR OIL JETS

VACUUM PUMP

TO CONSTANT-SPEED UNIT

C.S.U. FILTER

FROM RELIEF VALVE

MAIN PRESSURE VALVE

LOW PRESSURE VALVE

DRAIN TO CRANKCASE

TO WHEELCASE

FROM PRESSURE PUMP

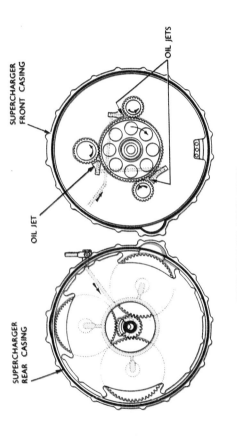

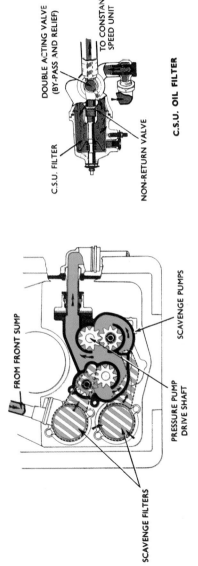

Oil circulation diagram

PHOTOGRAPHS OF MERLIN 100 SERIES

Merlin 113 – Starboard Side – Mosquito Installation

Merlin 114 – Starboard Side – Mosquito Installation

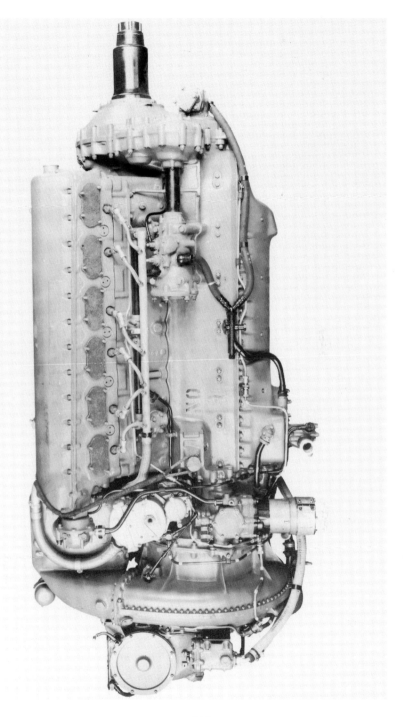

Merlin 130 – Starboard Side – Hornet Installation

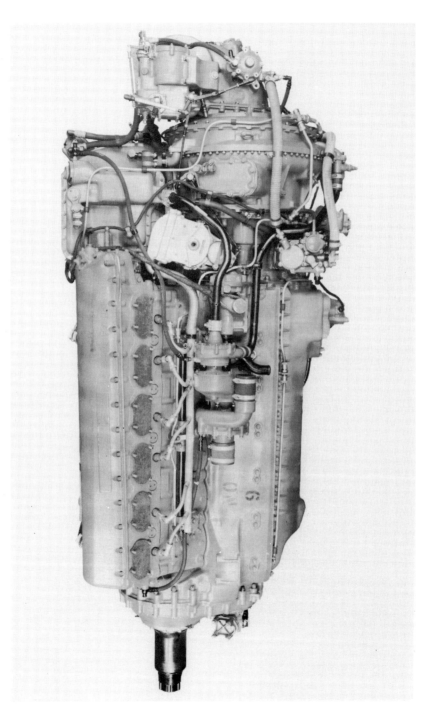

Merlin 130 – Port Side – Hornet Installation

Merlin 131 – Starboard Side – Hornet Installation

Merlin 140 – Starboard Side – Sturgeon Installation

Merlin 140 – Port Side – Sturgeon Installation

The Rolls-Royce
MERLIN Mk. 150

Merlin 150 – Later to become the Merlin 620 – an early Tony Dunwell Illustration